THE HUMAN DNA MANUAL

Contents

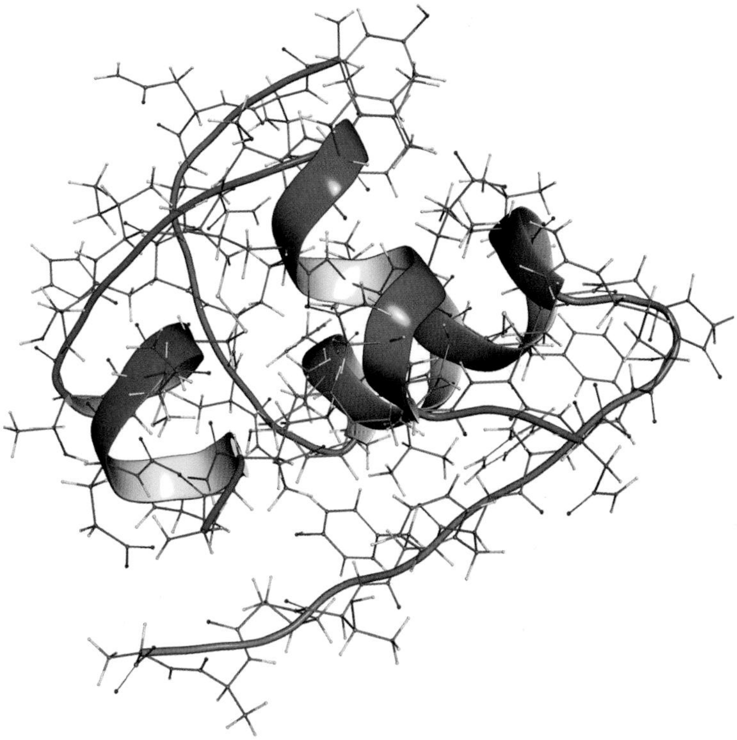

CHAPTER 1

The history of DNA discovery

It is impossible to tell the story of DNA discovery without first acknowledging the contribution of many pioneering scientists whose brilliant work set the scene. Through their discoveries, we know today that DNA takes the form of a double helix – crudely, two strands of genetic data that wind around each other and are held together by complementary links. DNA is found inside the cell nucleus, supercoiled into structures called chromosomes. It contains genes, the units of heredity; it holds the code to our very being. But this knowledge did not come to us in an instant. It is the result of hundreds of years of discovery, and the learning still continues today. To understand the history of DNA and its significance, we must first look inside our cells, where our DNA lives and operates.

Cells

Human beings have trillions of cells, the tiny units of life that make up our bodies' tissues. There are 200 different cell types, all with distinct characteristics that relate to their function. At 25 micrometres, half the width of a human hair, cells are so small that we can see them only under a microscope.

In 1665, Robert Hooke (1635–1703), an English physicist, discovered that the cell itself is made up of even smaller sub-units called organelles. These subcellular structures work together as a biological factory inside the cell, functioning in many different ways to keep us alive.

Organelles, cytoplasm and nuclei

Floating around in a gel-like filling, called cytoplasm, organelles are visible in any detail only with the help of a device called a scanning electron microscope, which can magnify images up to 500,000 times their original size. Cytoplasm is a mixture of water, salts and proteins and is home to the manufacturing departments of the cell – organelles known as the endoplasmic reticulum and Golgi apparatus – and the cell's energy supply, the mitochondria. It is also home to the nucleus. Although not strictly an organelle, the nucleus is the executive hub of the cell, the driver behind nearly everything the cell does.

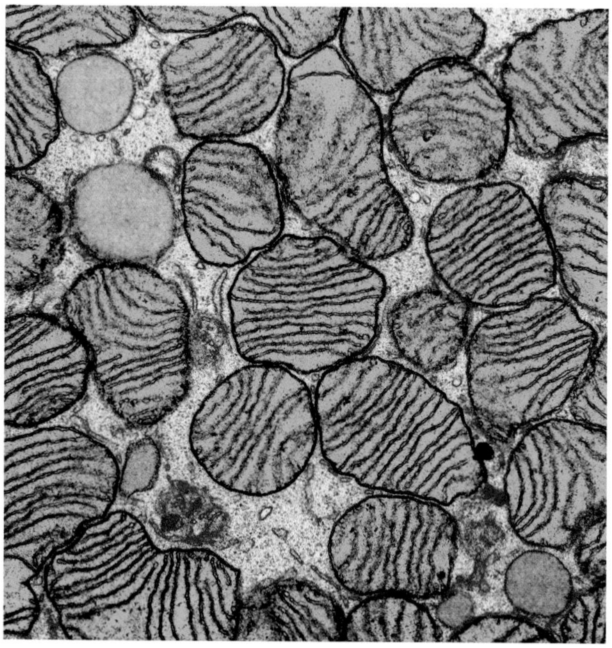

▲ **Multiple mitochondria in a cell, captured in fine detail by an electron microscope.**

A human cell

Inside a human cell lie the organelles. At the heart of the cell is the nucleus, brimming with DNA. In the surrounding cytoplasm lie the endoplasmic reticulum and Golgi apparatus, the protein factories, and the mitochondria that generate the cell's energy.

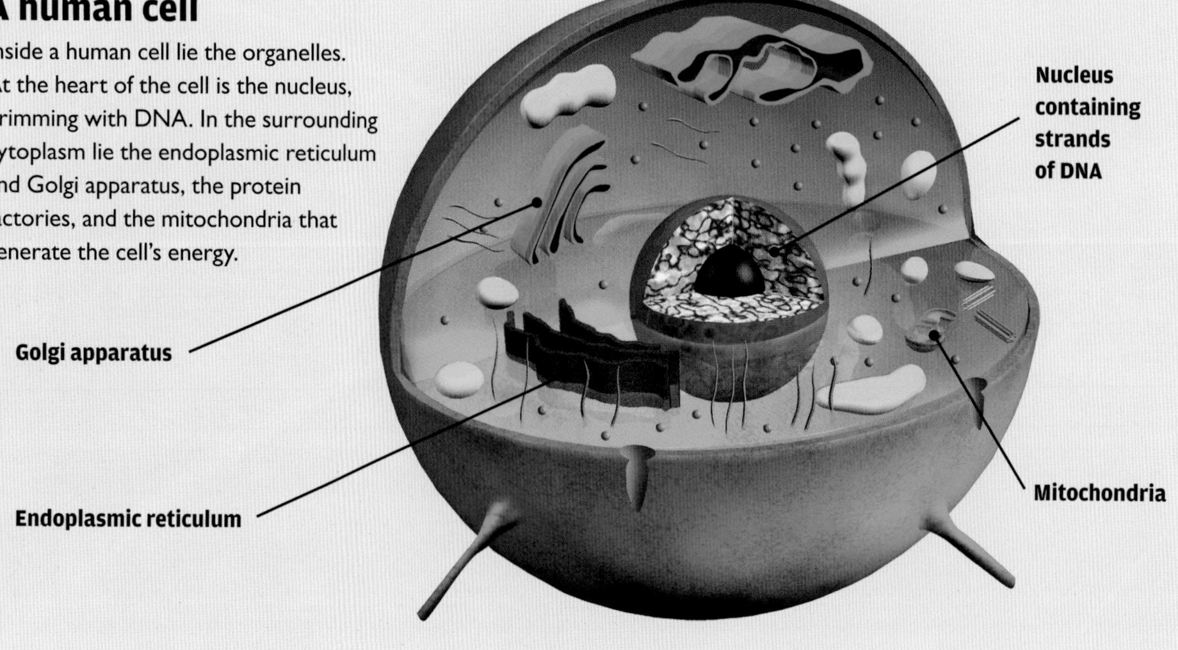

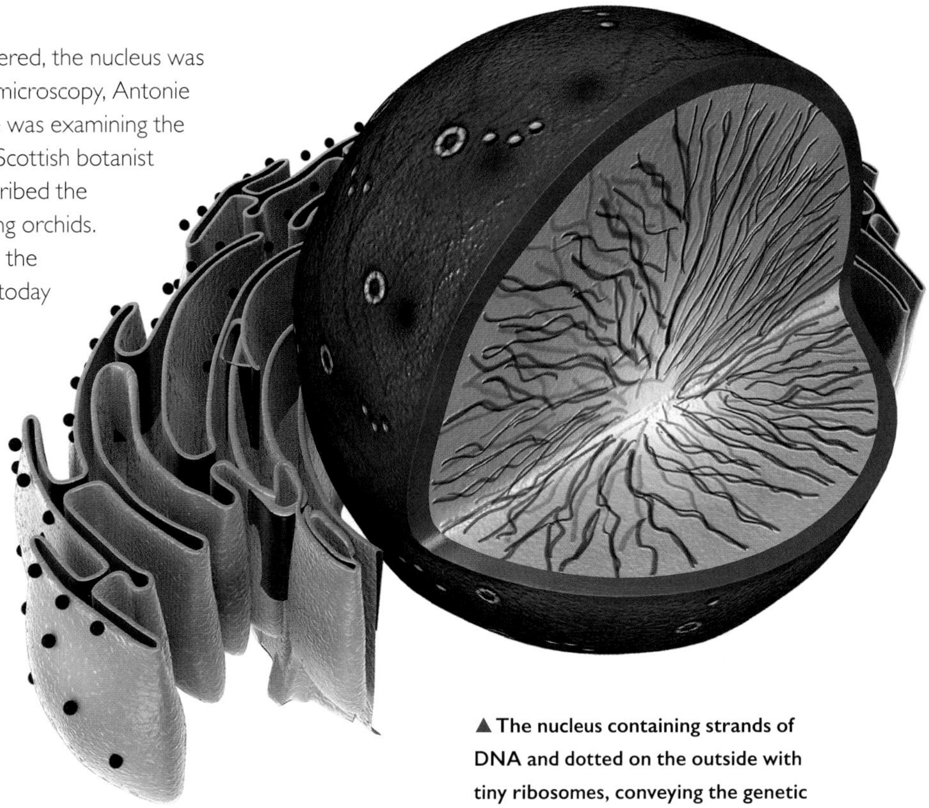

▲ **The nucleus containing strands of DNA and dotted on the outside with tiny ribosomes, conveying the genetic code to the endoplasmic reticulum.**

Nucleus

The first sub-unit of the cell to be discovered, the nucleus was observed in 1719 by revered pioneer of microscopy, Antonie van Leeuwenhoek (1632–1723), while he was examining the blood cells of the salmon. Later, in 1831, Scottish botanist Robert Brown (1773–1858) further described the nucleus, which he observed while studying orchids. At the time, neither scientist understood the true significance of their discoveries, but today we know that the nucleus is the cellular home of DNA, providing every cell in the body with the blueprint necessary for life through the genetic code.

The nucleus has a covering, the nuclear membrane, which separates it from the rest of the cell. The surface of this nuclear membrane is dotted with ribosomes, like tiny studs. They convey the messages held within the genetic code to other organelles, including the endoplasmic reticulum, where proteins are made.

Endoplasmic reticulum

In the factory house of the cell, the endoplasmic reticulum (ER) assembles the proteins, the building blocks of life. For this reason, it has to lie close to the nucleus – which houses the instruction manual. The ER is a network of tunnels where proteins are constructed, tweaked and modified, so that they are fit for purpose. The ER then communicates directly with another part of the cell, the Golgi apparatus, for the next part of the process: applying the finishing touches and getting the proteins to where they need to go.

Golgi apparatus

After they have been made in the ER, proteins are transported to another, connecting organelle, the Golgi apparatus. Here, the proteins are finessed and packaged up, then dispatched to their final destination – a bit like a parcel delivery service.

Mitochondria

Nothing happens in our body without fuel. In a cell, another organelle – the mitochondrion (plural mitochondria) – is the power station, the provider of the energy each cell needs to produce maximum output. There are between 1,000 and 2,000 mitochondria contained within the cytoplasm. Cells with a higher energy requirement, such as those within the liver and heart muscle, have a higher number of mitochondria.

mtDNA

Each mitochondrion contains its own DNA – a single, circular strand that is distinct from the DNA inside the cell nucleus. It also has most of the cellular machinery that it needs to live and function independently. In this respect, mitochondria resemble single-celled organisms (microbes), such as bacteria and viruses. Microbes invade the cells of the host they are infecting, then rapidly get to work, using their own substations to replicate and move on to the next host before the cells' protective machinery recognises them as foreign and destroys them.

Some sources suggest that mitochondria were originally invading bacteria that somehow, over millions of years of evolution, escaped our cells' seek-and-destroy mechanism. The theory goes that the mitochondria's predecessors lived in harmony with our cells, providing them with essential energy and the ability to self-replicate. As a result, the body adopted these invaders as card-carrying organelles in their own right.

Chromosomes

Chromosomes are structures found inside the cell nucleus. They are, in fact, tightly coiled strands of DNA, twisted around special proteins, called histones.

We should each have 46 chromosomes in total: 23 originate from our mother and 23 come from our father. Chromosomes can be further subdivided into types, called autosomes and sex chromosomes. Although chromosomes are tiny structures located inside the cell nucleus, we can see them under a microscope in a laboratory using such techniques as chromosomal karyotyping.

Karyotype analysis

The karyotype analysis, or karyotyping, is a test that counts the number, assesses the size, and studies the shape of a person's chromosomes. The resulting description is called their karyotype – usually 46,XX or 46,XY, the normal female and male complements, respectively. An abnormal result can be the cause of certain genetic conditions, such as Down syndrome (also known as trisomy 21), which occurs when a person has an extra copy of chromosome 21, giving all or some of their cells 47 instead of 46 chromosomes altogether.

Scientists who study chromosomes are called cytogeneticists. They use a standard blood sample and look specifically at the nucleus of white blood cells during the metaphase of their life cycle – the phase when the chromosomes are most obvious. (Interestingly, red blood cells do not have a nucleus because they eject it from the cell as they mature.) Cytogeneticists apply a special stain, Giemsa, to create a distinctive pattern of stripes, known as G-bands, that is unique to each type of chromosome. Each chromosome has a slightly different structure, based around a short arm (p-arm) and a long arm (q-arm). Using the banding pattern and the p and q arms as a starting point, cytogeneticists can check for chromosomal abnormalities.

Through the karyotyping technique, we know that there are different types of chromosomes, the autosomes (chromosomes 1 to 22) and the sex chromosomes (X and Y). This is an important distinction because it is also reflected in the genes that they contain within the DNA.

Autosomes

These are the chromosomes that are identical in both males and females. They are numbered 1 to 22, based on their size, with

Human female karotype

These chromosomes have been stained to reveal G-bands, stripes that are unique to each chromosome, allowing them to be arranged in this pattern. We have 46 chromosomes. A female has two X chromosomes (46,XX).

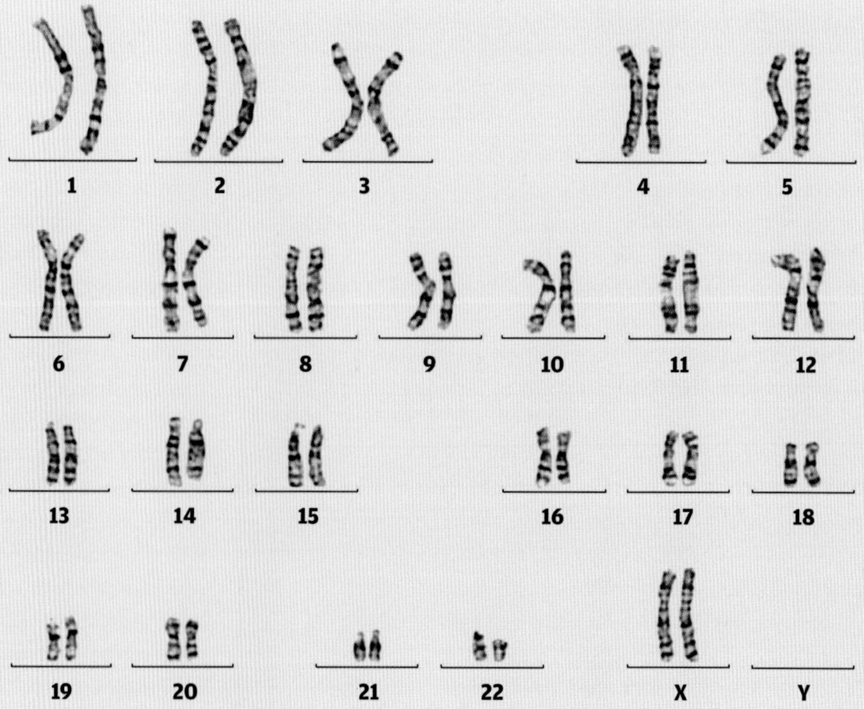

number 1 being the longest with the most genes. They become progressively smaller in size, so chromosome 22 is the shortest and contains the fewest genes. All 22 autosomes are exactly the same in males and females because we inherit a full set from our mother and another full set from our father. Each set is passed down to us in the egg and the sperm that created us. This is the way we inherit genes from both parents.

Contrary to popular belief, individual chromosomes are not wholly responsible for specific body organs. For example, chromosome 10 is not the heart chromosome, nor is 19 in charge of the brain. Xhromosomes distribute our genes in an apparently random way. This makes predicting the effects of a chromosome abnormality difficult, unless it is an abnormality that happens over and over again, as with Down syndrome.

Sex chromosomes

So, if the autosome pairs give us 44 chromosomes, we still need a further two chromosomes (another pair) to give us the full complement of 46. This is where the sex chromosomes, X and Y, come in. Females have two X chromosomes (46,XX) and males have one X and one Y chromosome (46,XY). A mother always gives her baby an X chromosome. When a father passes on his X chromosome, too, the baby is a girl; when he gives his Y chromosome, the baby is a boy.

The X chromosome contains about 800 genes. In comparison, the Y has only about 70, including the so-called

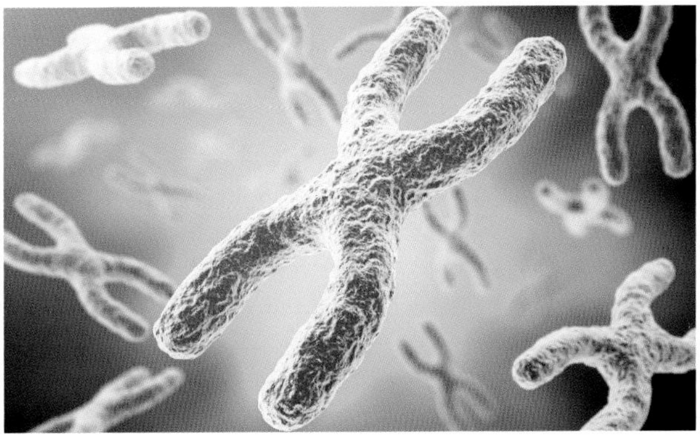

sex-determining genes that make someone with a Y male. So how do males function with only one X chromosome if it carries so many important genes? English geneticist Mary Lyon (1925–2014) discovered that female embryos actually switch off one of their X chromosomes and also function using only one. This switching-off process is called Lyonisation in Mary's honour. Which X – the one from the mother or the one from the father – is inactivated is a random event in every cell.

So, it's perfectly possible to live with only one working copy of the X chromosome. Not only is that the norm in males, but girls living with a condition called Turner syndrome (45,X – where the second sex chromosome is missing) are also perfectly healthy. However, it is not possible to live with only a Y chromosome. Pregnancies where this happens result in early miscarriage.

Human male karyotype

Males have a similar karotype to females, sharing the same autosomes (chromosomes 1-22), but differing in their sex chromosomes. Males have an X and a Y chromosomes (46,XY).

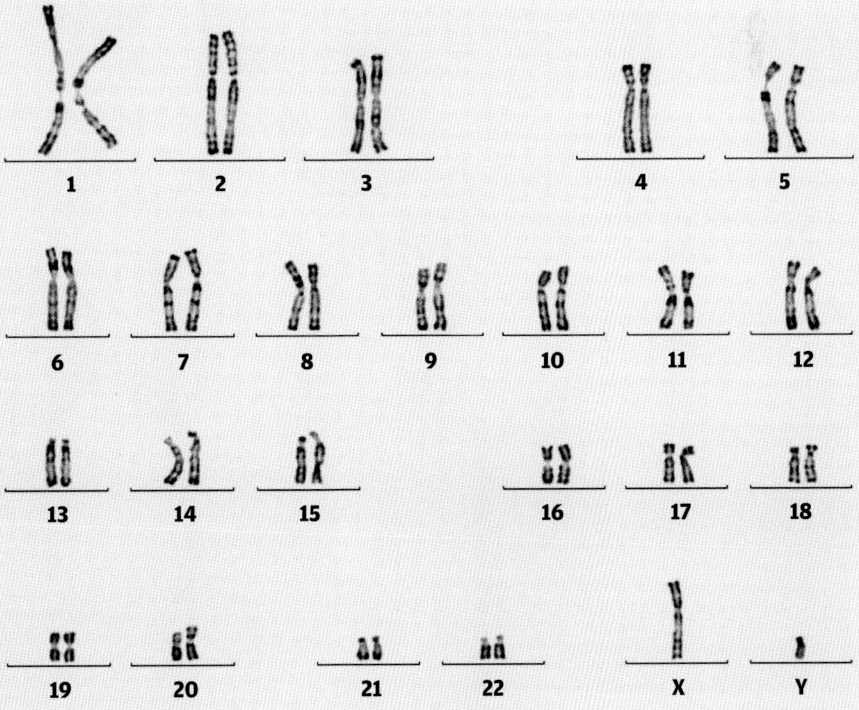

Cell division

From the moment the sperm fertilises the egg at conception, cells begin to divide, creating new cells from existing ones. The cells of an embryo divide rapidly, and so the embryo quickly grows to become a foetus, a tiny human, complete with internal organs and body systems comprising tissues made up of specialist cell types. It's the genetic code that controls these early developmental processes. After birth, there is another period of rapid growth, new cells constantly forming by repeated cell division. This system of dividing and multiplying occurs in two ways: mitosis and meiosis.

Mitosis

In the process of mitosis, two daughter cells form from one mother cell. The new cells are exact replicas of the original, even down to the DNA inside the chromosomes. The DNA in the new cells must be a faithful copy every time, in order to ensure that no mistakes are introduced that might change the genetic code.

Mitosis occurs throughout life – old cells dividing to create new ones, continually renewing and replenishing the stocks. Each of the two daughter cells contains the same number of chromosomes as the mother cell. The new pair of cells is known as a diploid, meaning two matching sets (of 46 chromosomes altogether).

Meiosis

Also known as 'reduction division', meiosis is different from mitosis because its purpose is to halve the number of chromosomes in the equivalent of the daughter cells. This is essential for healthy 'gametes', the cells essential for reproduction – our eggs or sperm. After two separate cell divisions, the finished product has only one chromosome set instead of two, 1–22, along with one of the sex chromosomes, either X or Y, giving a total of 23 chromosomes altogether (a haploid). Before meiosis occurs, chromosomes belonging to the same pair line up alongside each other and swap material in a genetic reshuffle. This ensures that the chromosomes passed on from parent to child are no longer the same as the original, but a new combination – a mixture of the generation before that gives a new and unique code to create a new and unique person.

Mitosis

Mitosis is the process by which a cell divides, creating two new identical cells by nuclear division. The cell proceeds through a number of different phases to replicate the chromosomes and separate the new copies from the original ones, moving them apart to different poles of the cell. It then pinches off in the middle to form two new cells, each containing the full set of chromosomes.

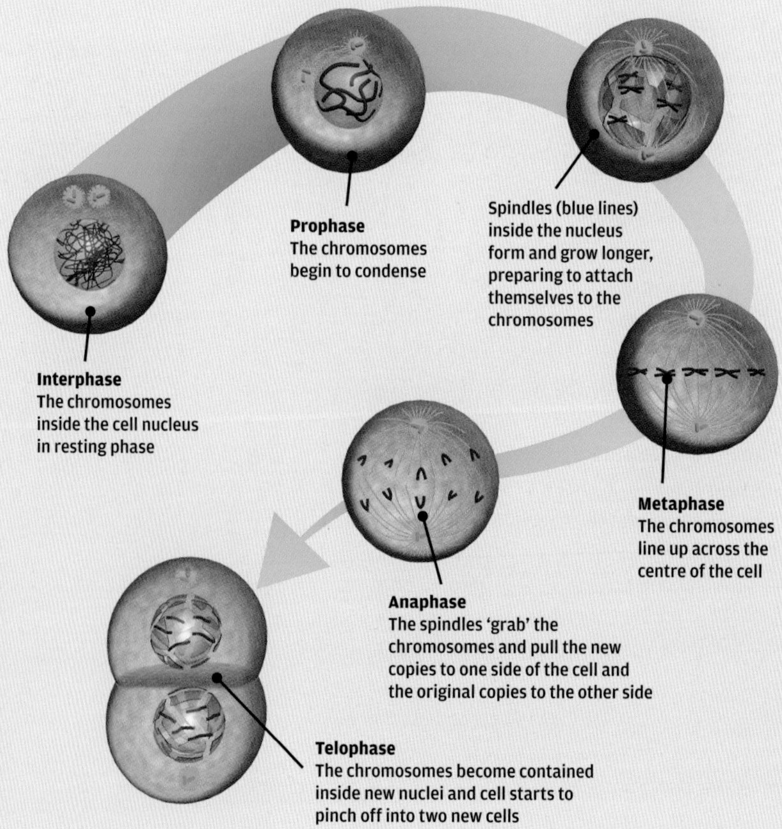

▼ The final stage of cell division, creating two daughter cells, each identical copies of the original.

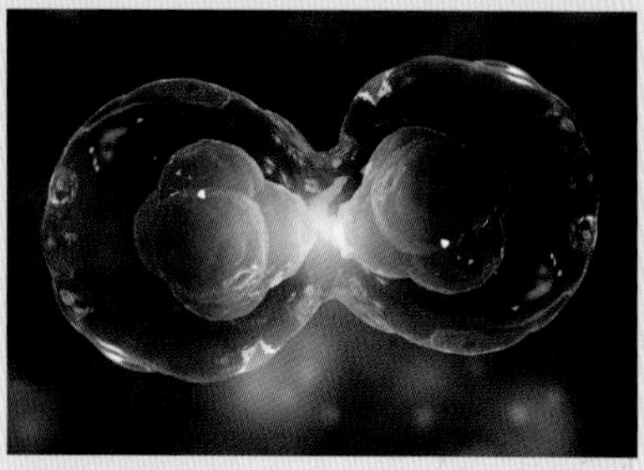

Meiosis

Meiosis is also known as reduction division because the number of chromosomes in the cell is halved, from 46 to 23. This is the process required to make eggs and sperm. Just as with mitosis, the chromosomes separate and move to different poles of the cell, but this time they do not copy themselves first. Instead, each combines with its opposite number to undergo crossover and the new chromosome versions finally end up in the egg or sperm

1 **Meiosis 1** The chromosomes duplicate

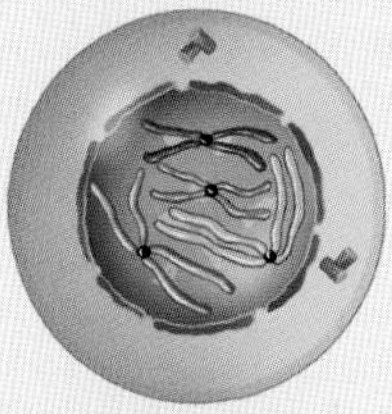

2 Homologous or same number chromosomes align with each other and crossover genetic material

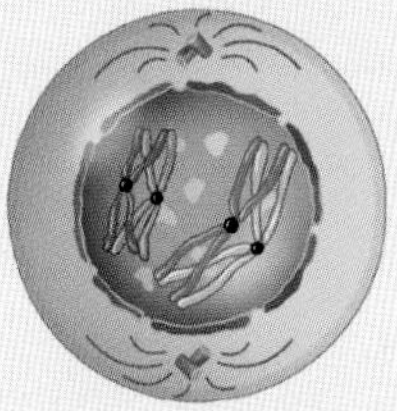

3 Spindles attach to the chromosomes as they line up across the centre of the cell and start to pull them apart

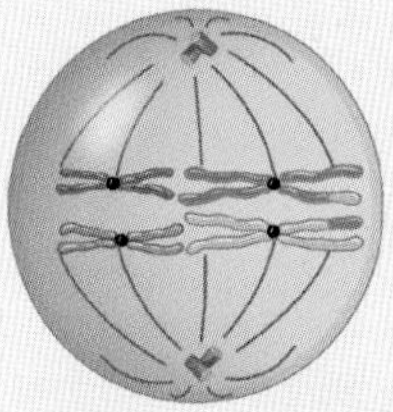

4 The chromosomes are pulled by the spindles to opposite poles of the cell so that two new daughter cells can form

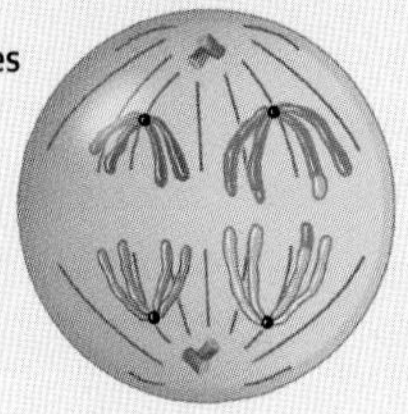

5 **Meiosis 2** In the two new daughter cells, new spindles form and attach to the separate arms of the chromosomes (chromatids)

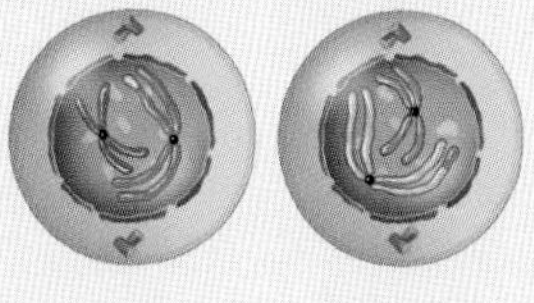

6 The spindles pull the chromatids apart to opposite poles of the cells, which then split forming two new cells, four new cells in total from the original cell at the start of meiosis 1

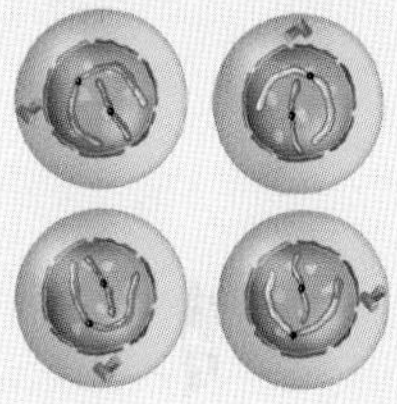

Trisomy 21

The process of meiosis is not fail-safe. In spite of the nucleus running safety checks, mistakes, known as chromosomal imbalances, can happen. Arguably, the best-known example of a condition resulting from chromosomal imbalance is Down syndrome. In a person with Down syndrome, instead of there being two copies of chromosome 21 (disomy), there are three (trisomy). This is because an error occurred during the meiosis that created the cells of the egg. We know that for that particular egg, the chromosome-21 pair did not separate and reduce, and the resulting egg had two chromosome 21s rather than just one. If the ovary releases the egg containing those two copies of chromosome 21 and a sperm (which contains one copy itself) fertilises it, the embryo ends up with three copies altogether – trisomy 21.

Rearrangements, changes in the structure, can occur at the meiosis stage too. Among these rearrangements are translocations, in which sections of chromosome material swap places with one another. If this does not alter the amount of chromosomal material present in any way, it is called a balanced translocation. However, if any of the material is missing or extra, it is an unbalanced chromosome translocation, which usually has a damaging effect. In turn, this causes either miscarriage or the birth of a baby with congenital abnormalities, significant growth disorder or delayed cognitive development. Routine screening using blood tests and ultrasound scans during pregnancy are designed to detect unexpected events like this.

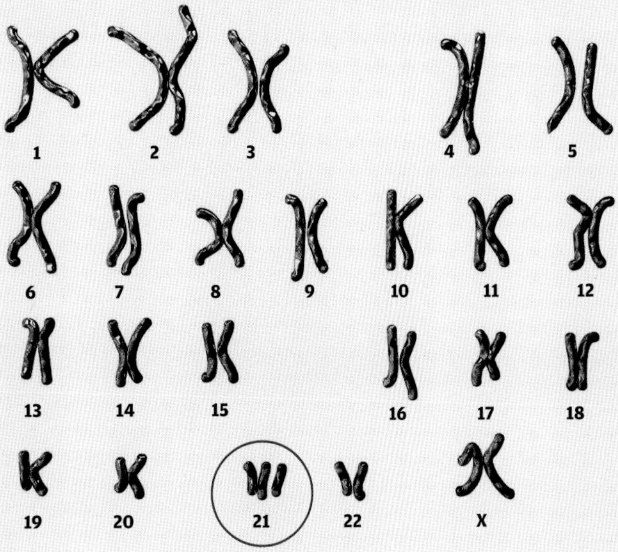

Heredity

Heredity is the passing on of traits from one generation to the next, a phenomenon described in great detail by the father of modern genetics, Gregor Mendel (1822–1884), although the significance of his work was not realised until after his death. Mendel was a friar from the Augustinian Abbey of St Thomas in Brno, Czech Republic. There he grew and studied the common pea plants, *Pisum sativum*. He cross-bred the plants in different combinations, observing and recording several different characteristics of the plants, such as seed colour and shape, stem length, position of flowers on the stem and the colour of the pea pods. He did this in such intricate detail across several generations that he was able to determine that the transmitted traits he observed resembled either one parent plant or the other, and not an amalgamation of the two, as scientists before him had assumed. For example, he noticed that the seeds were either wrinkled or smooth, passed on as the same trait, not a blended combination. What's more, he observed that some traits were passed down in a different pattern to other traits.

Over a period of eight years, he conducted experiments on his pea plants aiming to determine just how this happens. He published his findings in 1865, when he concluded that there were three 'laws' that determine how traits are inherited: the law of dominance, the law of segregation and the law of independent assortment.

The law of dominance

In the law of dominance, one trait (the dominant trait) suppresses the other (the recessive or suppressed trait). When Mendel crossed pure-bred plants that had red flowers with pure-bred white-flowered plants, all the offspring plants had red flowers, confirming that red flowers were the dominant trait, while white flowers were recessive.

Mendelian inheritance

Mendel used yellow and green peas in his experiments, demonstrating that crossbreeding generated only yellow plants in the first generation (dominance), but green and yellow peas in a ratio of 1 to 4 when these hybrids were crossbred. Genetic variation was also evident in other plant characteristic that display dominant (R) and recessive (r) traits, demonstrated through crossbreeding of hybrid generations.

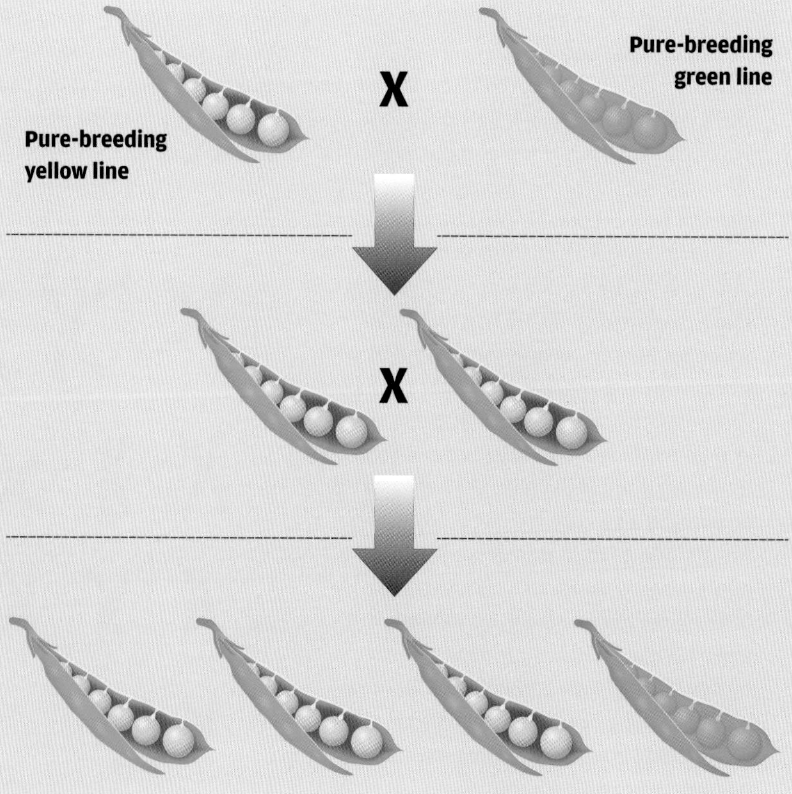

The law of segregation

According to the law of segregation, a parent passes on only one version – or allele – of a gene to the offspring. In the previous example that's the trait of the red flower or the white flower. That an allele producing one trait does not influence an allele resulting in another, unrelated trait underpins the third of his principles, the law of independent assortment.

The law of independent assortment

In one of his cross-breeding or hybrid experiments, Mendel looked closely at red and white flowers. In the true-bred plants he created through cross-fertilisation, he knew that in one set of plants all the alleles were for red and, in the other, all the alleles were for white. He assigned these plants the abbreviations RR (all red) and rr (all white), where an upper-case letter denotes a dominant gene and a lower-case letter denotes a recessive gene.

When he crossed these pure-bred plants together, he observed that the number of red flowers to white in the next generation was three to one in favour of red. He noted this using the following system: RR (the plant has inherited two dominant R alleles), Rr (the plant has inherited one

▲ **Gregor Johann Mendel.**

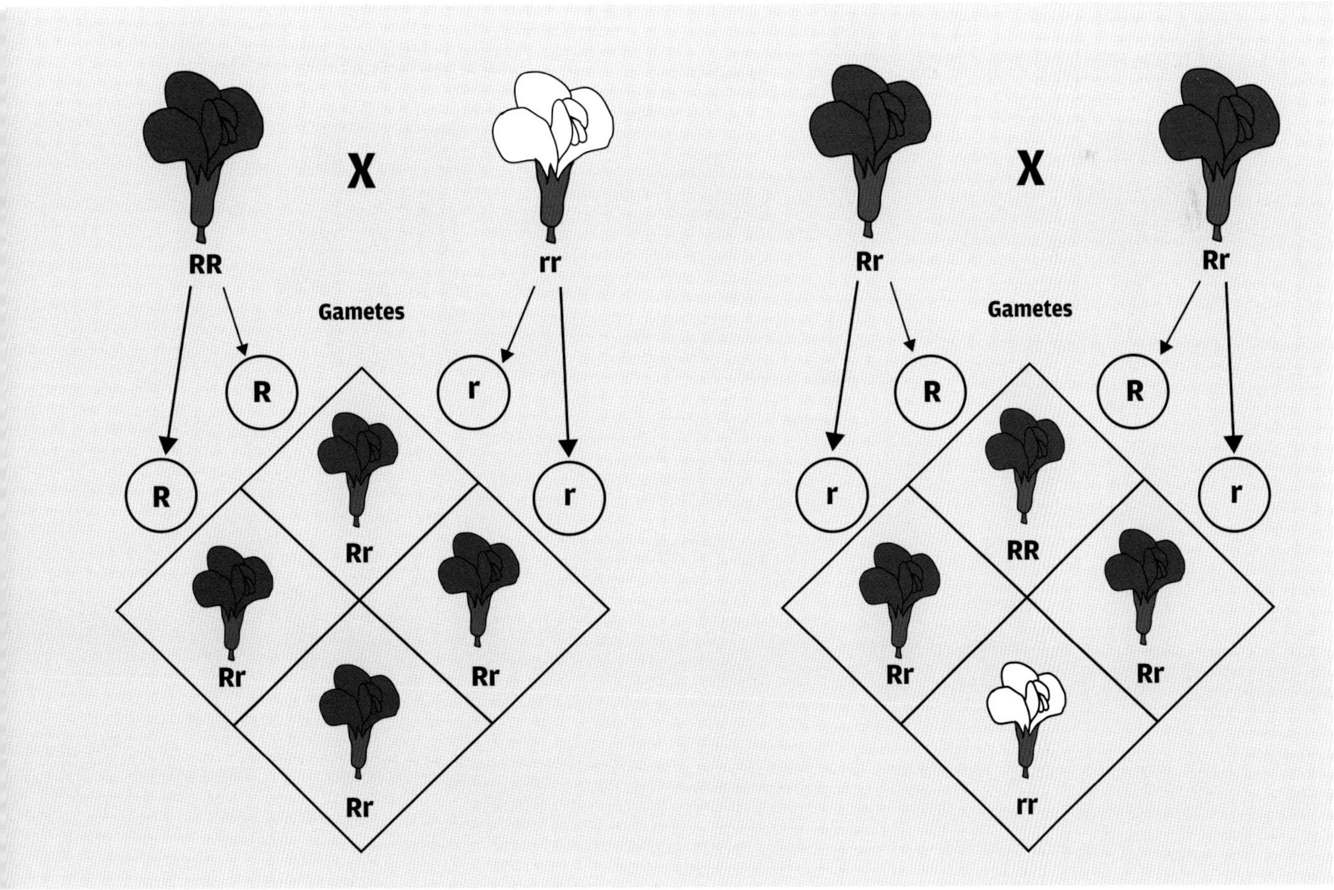

dominant allele and one recessive), rR (the plant has inherited one recessive allele and one dominant) and rr (the plant has inherited two recessive alleles). Only rr flowers, uninfluenced by the R allele, were showing up as white.

We can depict this experiment as a Punnet square, named after Reginald C Punnett (1875–1967), the biologist who first devised this visual representation. It shows that red dominates over white (law one, dominance), that white and red traits are passed on as separate traits (law two, segregation) and that the transmission of one trait to the next generation is not influenced by another allele, such as wrinkling of the seed (law three, independent assortment).

Gametes and zygosity

From Mendel's work it becomes clear that there must be cells specifically designed to conduct heredity to the next generation. We call these the gametes. In humans, they are the eggs and sperm. Which combination of alleles occurs in the offspring, RR, Rr or rr, is termed zygosity. Where the two alleles are the same, as with RR and rr, zygosity is the same, or homozygous. For Rr, where the alleles are different, they are heterozygous.

Inheritance patterns

Mendel's work lives on today in our understanding of the different patterns of inheritance. We now know that the unit of heredity that he observed to convey a trait from one generation to the next is in fact a gene. Today, geneticists construct pedigrees and look for clues that might reveal how a particular trait or disease is being transmitted, either autosomal or sex-linked (depending where on the chromosomes they find the gene) and dominant or recessive (relating to the suppressing power the trait has over the other allele).

There are four Mendelian inheritance patterns: autosomal dominant and autosomal recessive and X-linked dominant and X-linked recessive. As the mitochondria contain DNA too, we also have mitochondrial inheritance and, in theory, through Y-linked inheritance, a father can pass Y-chromosome traits on to his son, meaning there are eight different patterns of inheritance.

▶ In humans, the gametes are the egg and sperm, the way we pass our genes onto our offspring.

▼ Patterns of inheritance of autosomal and sex linked genes

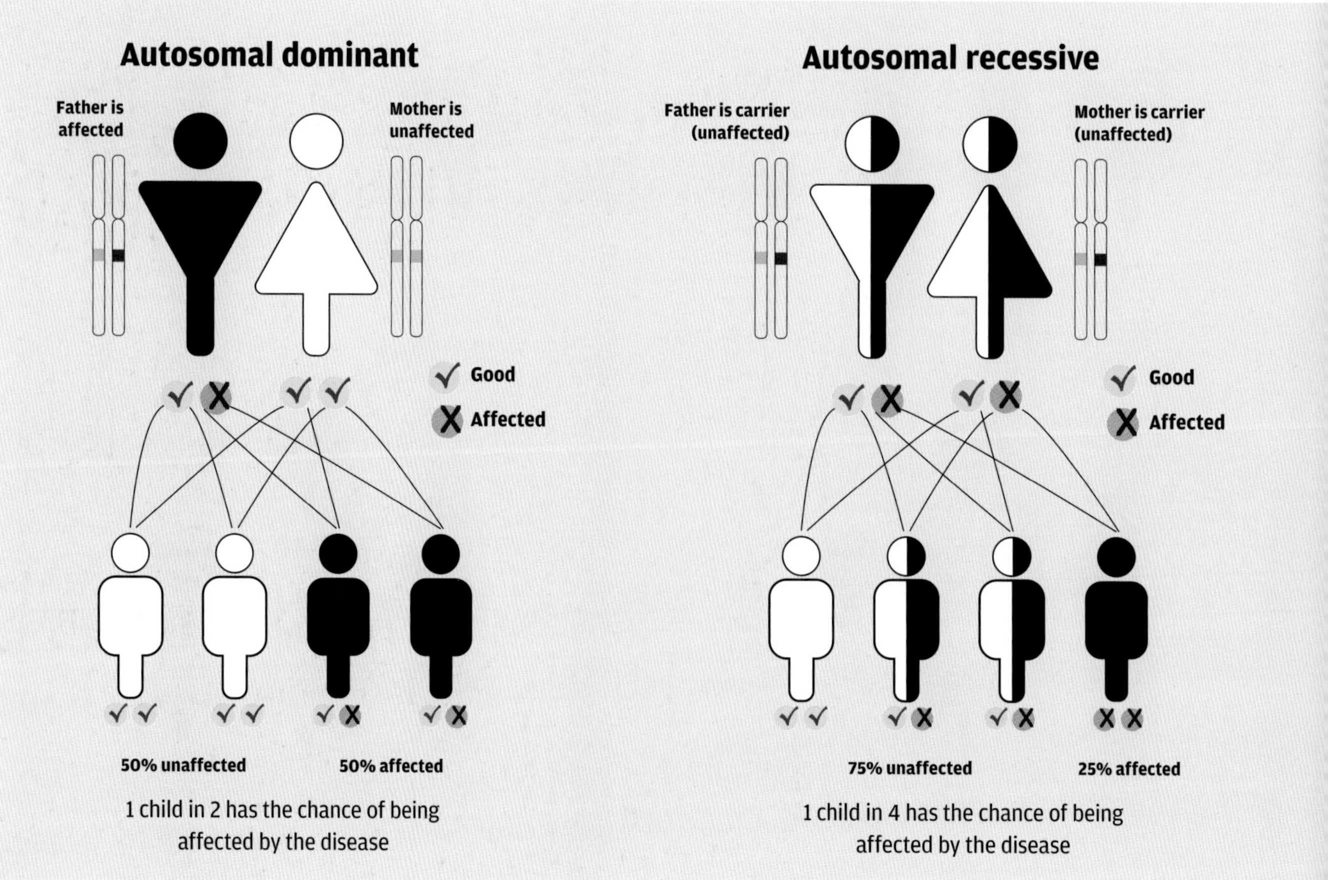

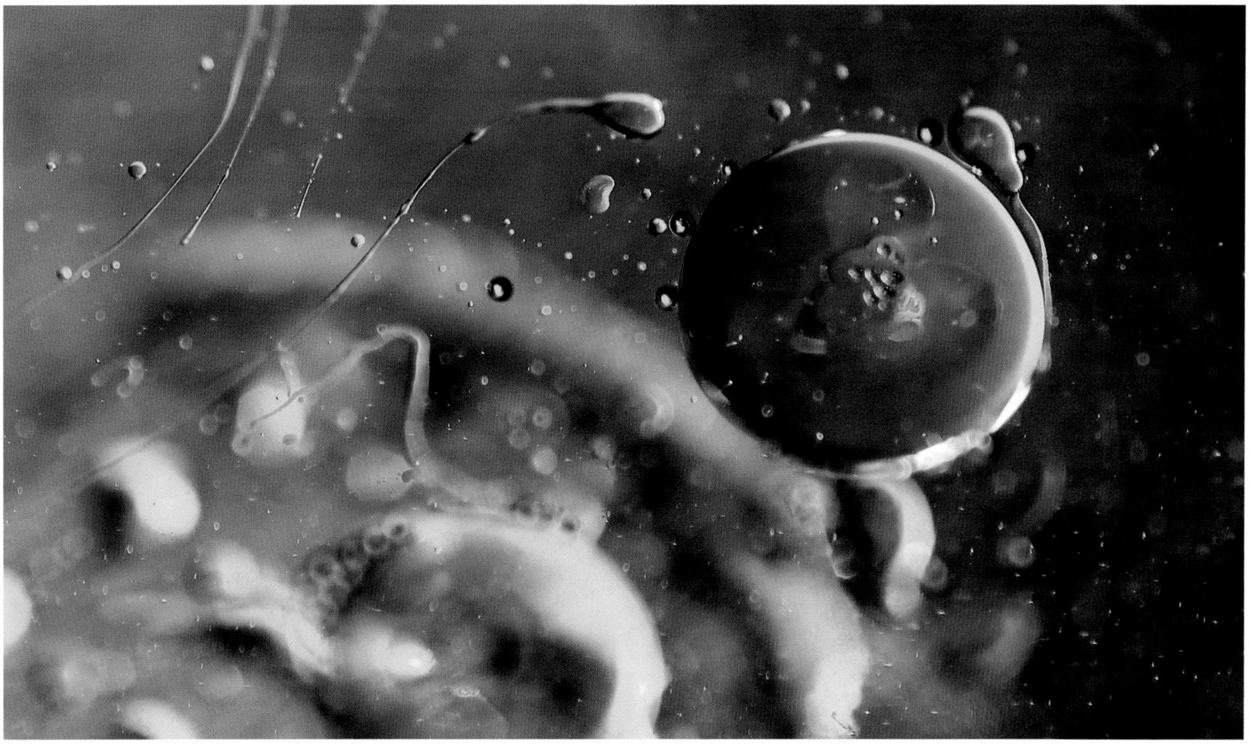

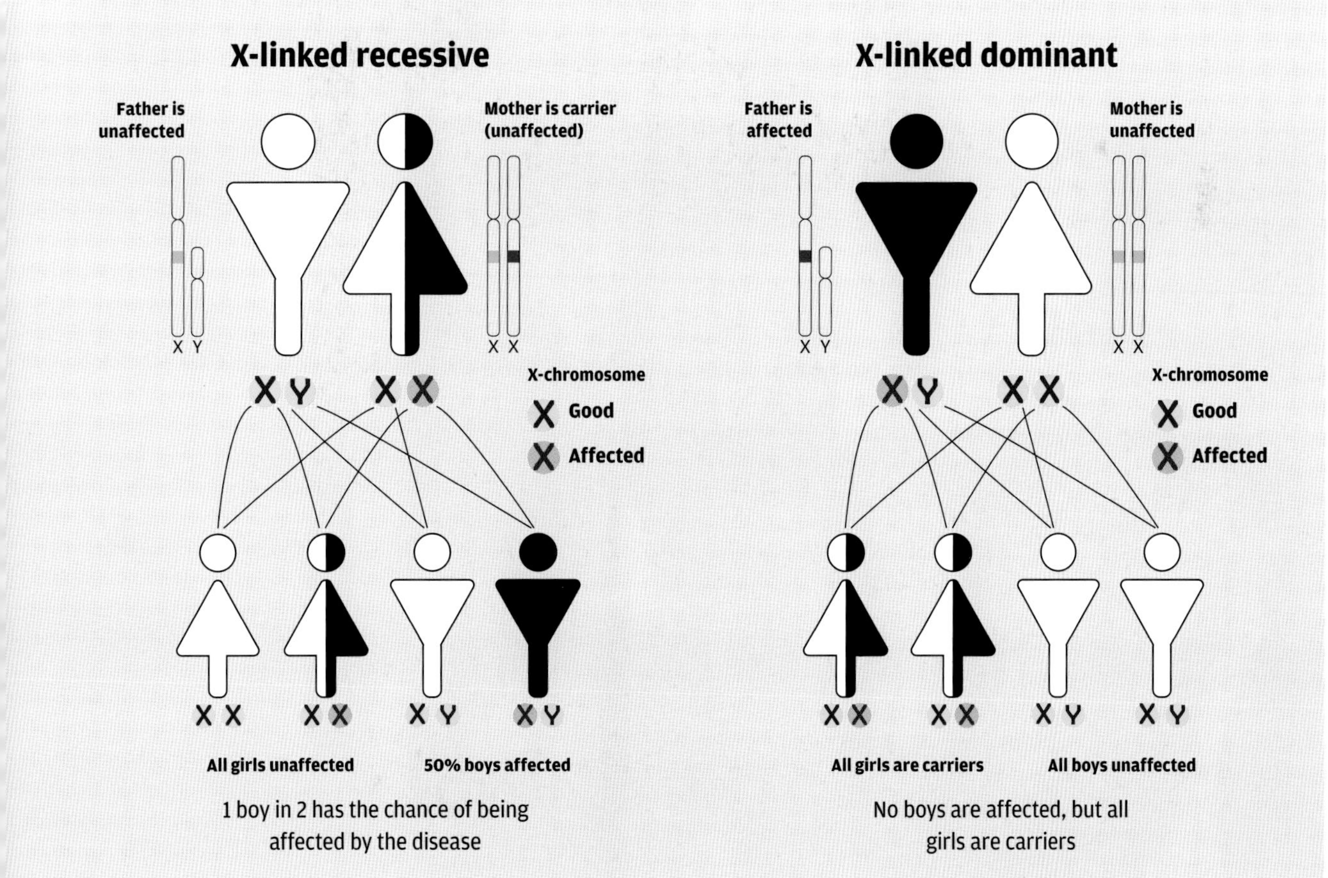
X-linked recessive
Father is unaffected
Mother is carrier (unaffected)
X Y
X X
X Y
X X
X-chromosome
Good
Affected
X X
X X
X Y
X Y
All girls unaffected
50% boys affected
1 boy in 2 has the chance of being affected by the disease
X-linked dominant
Father is affected
Mother is unaffected
X Y
X X
X Y
X X
X-chromosome
Good
Affected
X X
X X
X Y
X Y
All girls are carriers
All boys unaffected
No boys are affected, but all girls are carriers

Genes

Hugo de Vries (1848–1935), a Dutch botanist, correctly calculated that the heritable traits Mendel had observed were transmitted as particles. When he published his own findings in 1900, de Vries named these particles 'genes'. Then, in 1909, when English biologist William Bateson (1861–1926) translated Mendel's seminal publication *Versuche über Pflanzenhybriden* (*Experiments in Plant Hybridization*; 1865) into English, he called the study of heredity 'genetics'.

DNA

Initially, scientists believed that heredity was transmitted via proteins. That deoxyribose nucleic acid (DNA) is the heritable material found in genes was the work of three US-based scientists, Oswald Avery (1877–1955), Colin MacLeod (1909–1972) and Maclyn McCarty (1911–2005). They conducted experiments using a bacteria, *Streptococcus pneumonia*, showing that there was something present in a heat-killed virulent strain that could transform a non-virulent strain into a deadly one, the so called transforming principle. This 'something' turned out to be DNA and knocked on the head the theory that proteins were responsible for heredity.

▲ **Rosalind Franklin.**

◀ **James Watson, left, and Francis Crick, right.**

DNA molecule

DNA is a double helix made up of two strands each containing a backbone of sugar and phosphate, and interlinked by the nucleotide bases adenine (A), cytosine (C), guanine (G) and thymine (T), creating a double helix structure.

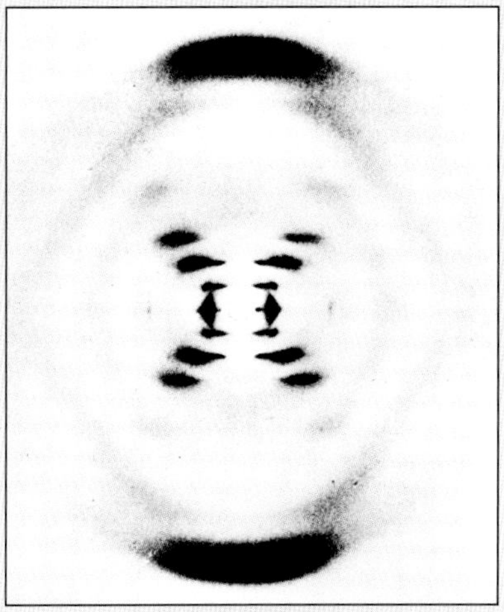

▲ Photo 51 was the picture generated by Xrays being scattered by DNA onto a photographic plate. It lead to the discovery that DNA is a double helix.

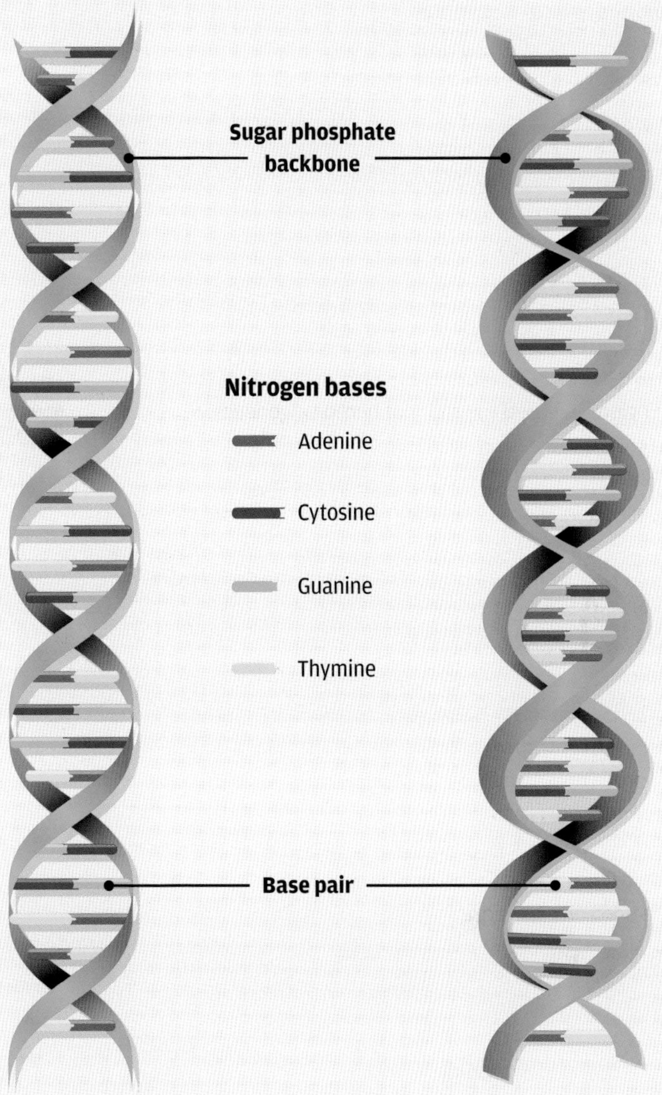

Nucleic acid

Around the time that Mendel was experimenting with pea plants, Friedrich Miescher (1844–1895) isolated a previously unknown acid from the nuclei of white blood cells. He called it nuclein. In 1889, German pathologist Richard Altmann (1852–1900) discovered that nuclein had a weakly acidic nature, which prompted him to adapt the name to nucleic acid.

Double helix

It wasn't until 1952 that we fully understood exactly how DNA conveys heredity. The secrets of DNA were uncovered, first, through the work of London-based Rosalind Franklin (1920–1958) and Maurice Wilkins (1916–2004), and subsequently in 1953 by Cambridge academics, James Watson (1928–) and Francis Crick (1906–2004).

Franklin and Wilkins used X-rays to generate an image of the DNA structure using a technique called diffraction. This was the prerequisite to Watson and Crick calculating that DNA is assembled as a helix made up of two strands. (It is this unique shape that gives DNA the ability to act as the unit of heredity.)

In the years following Watson and Crick's discovery, other scientists combined their knowledge and worked out the exact structure of the DNA molecule. They confirmed that it is a strand of nucleic acid containing nitrogen-rich bases called nucleotides, attached to a backbone made up of sugar and phosphate. Two of these strands wind round each other like a spiral ladder, the side rails constructed from a sugar-phosphate backbone and the rungs being the nucleotides linking with each other, creating the unique DNA double helix.

Sugar-phosphate backbone

The backbone of DNA contains the sugar deoxyribose and a chemical group called phosphate. The purpose of the backbone is to provide a scaffold for the nucleotide bases, so they can link together efficiently. Another type of nucleic acid also exists, ribonucleic acid (RNA), but its backbone is made of ribose rather than deoxyribose.

Base pairs

In order to understand the way in which DNA works as the hereditary material, we have to understand base pairs. There are four base pairs, called adenine (A), thymine (T), cytosine (C) and guanine (G). A and G, the purines, are a different shape to C and T, the pyrimidines. The two strands of DNA are complementary to each other, which means they always slot together in a certain way to make the double helix: A linked with T and C with G. These are the base pairs, and humans have 3 billion of them in their DNA.

DNA replication

When existing cells divide to make new cells, the DNA also has to replicate, so that the new cells contain a full set of genetic information, too. In 1958, Matthew Meselson (1930–) and Franklin Stahl (1929–), in their experiment that came to be known as the 'most beautiful experiment in biology', demonstrated that DNA does this through a process called semiconservative replication: one strand of DNA forms the template for the construction of a new one. In the daughter cells resulting from cell division, one strand of DNA originates from the mother cell (semiconservative) and the second strand is a copy of it (replication). DNA's ability to do this is held in the complementary nature of the nucleotide bases: A–T and C–G.

DNA replication

DNA replicates through a number of different steps designed to create a new strand of DNA that is identical to the original; its exact replica. The double helix unwinds, so each strand can act as a template to create a new strand. New bases that are complementary to the ones left exposed by the unzipping step attach them selves A to T and C to G, creating a new DNA strand.

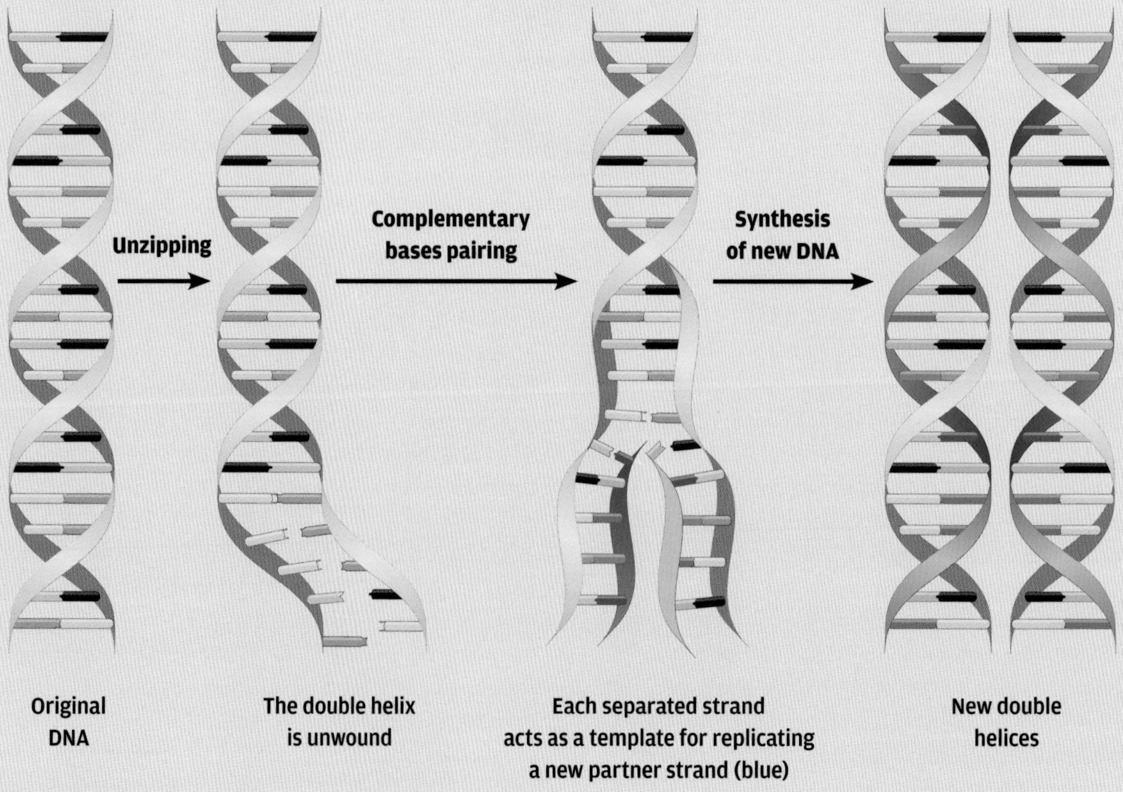

Cracking the genetic code

In 1961, Francis Crick and collaborators published their theory on how DNA is translated into proteins. In other words, how they had cracked the genetic code. There are three rules applied to the code.

The first is that it is universal. The four nucleotide bases are found in the DNA of every living thing. The same rules apply, no matter if you are a man, a banana, an orchid or a jellyfish. This rule applies to the codons and their matching amino acids, too.

The second rule is that the genetic code is non-overlapping. This means that three bases are read separately, one after the other, just like different words in a sentence.

Finally, there are 20 amino acids and four bases. If the codons were made up of two bases only, there would be only 16 possible combinations, which is not enough for 20 amino acids. Three-base codons offers the luxury of 64 possibilities, more than enough for the job, and so some amino acids are assigned more than one triplet, the so-called 'degenerate' code. Some of these triplet codes act as start-and-stop codons, showing where protein translation should begin and end.

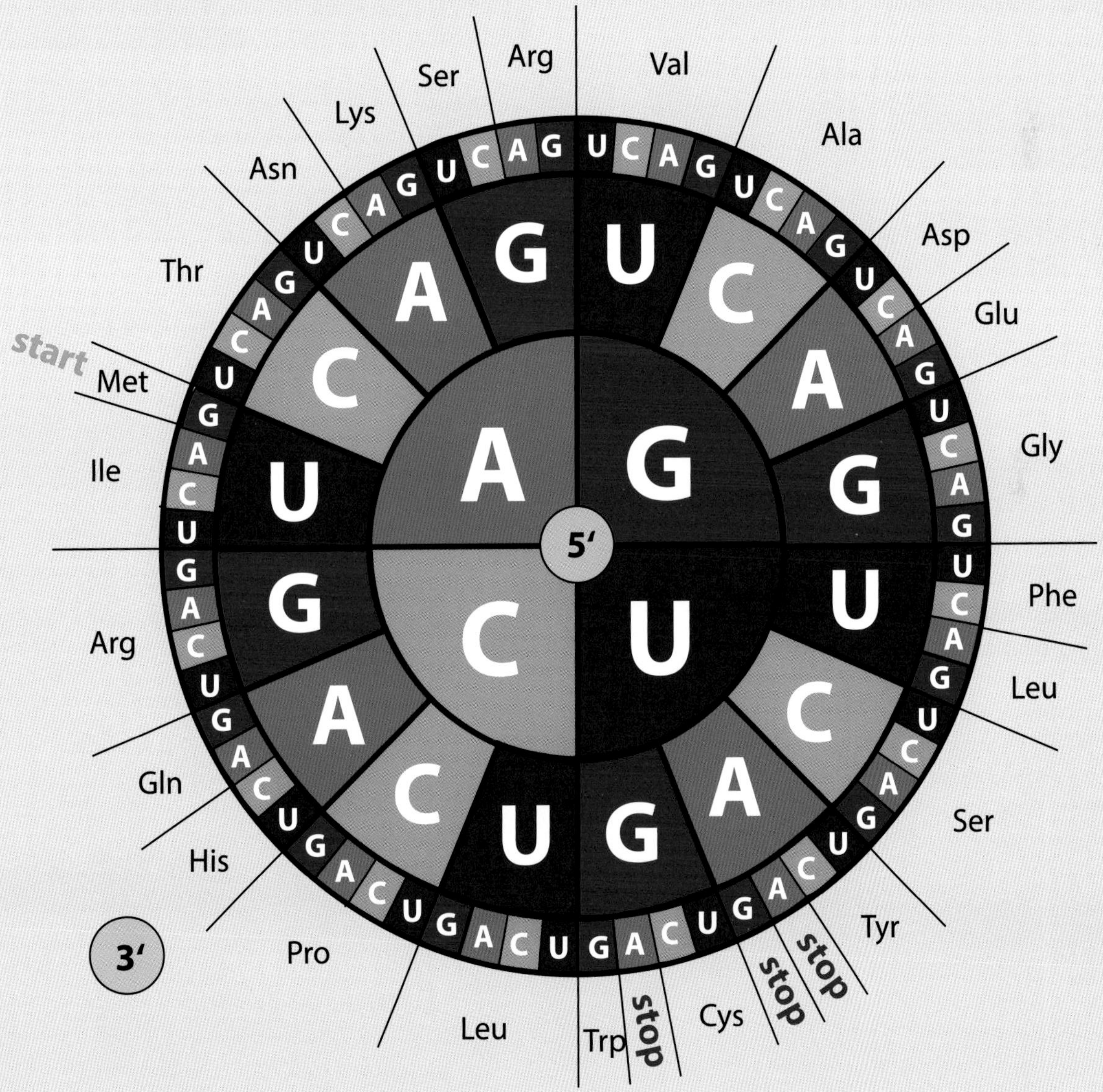

RNA and ribosomes: the tool box

With DNA firmly held as the molecule of heredity and proteins being the building blocks of life, scientific attention moved to the link between the two. How does protein formation actually happen?

Francis Crick hypothesised that there must be something to link DNA to protein formation. In 1957–1958, American biochemist Robert Holley (1922–1993), working with fellow biochemists Marshall Nirenberg (1927–2010) and Har Gobind Khorana (1922–2011), discovered that link: RNA.

RNA is similar to DNA, but with two important differences. The sugar in the backbone of RNA is ribose rather than deoxyribose, and instead of the nucleotide base T (thymine), RNA substitutes uracil (U), so A–U are complementary to each other (C and G stay the same).

Proteins are assembled when the messages hidden within the DNA are decoded in a two-step process: DNA to RNA, RNA to protein. Known as the 'central dogma of molecular biology', the process requires three different kinds of RNA – messenger RNA (mRNA), transfer RNA (tRNA) and ribosomal RNA (rRNA) – as well as ribosomes, to translate the genetic code into protein.

mRNA

In the first step to unlocking the DNA code, the double helix partially unzips, letting in RNA molecules so that they can bind to the single strands and make a run of mRNA. This unedited mRNA also contains sections of so-called 'junk', transcribed from the non-protein-making regions of the genes, called introns. Rather like ads in a magazine or newspaper article, introns contribute nothing to the relevant information and are best cut out or spliced, leaving just the sections key for translation into protein.

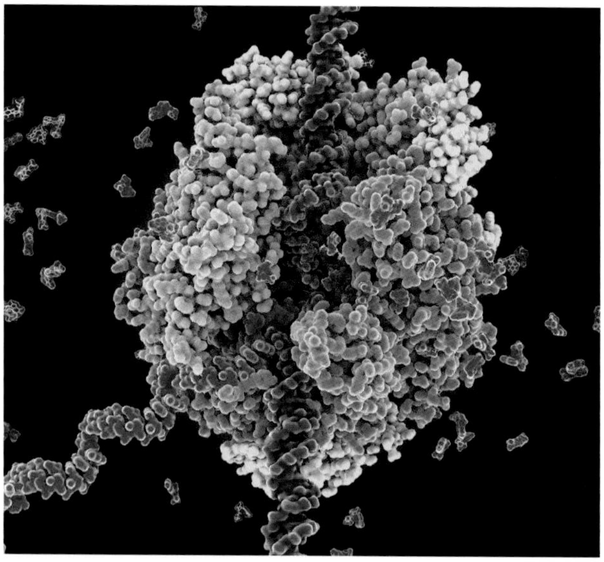

▲ RNA polymerase is composed of several proteins. It unwinds DNA strands (violet) and builds RNA (red) out of the nucleotides uridine, adenosine, cytidine and guanosine monophosphate.

The central dogma of molecular biology

The central dogma of molecular biology explains how proteins are derived from the genetic code. It is a three-step process in which DNA replicates and is transcribed into RNA, which is then translated into proteins.

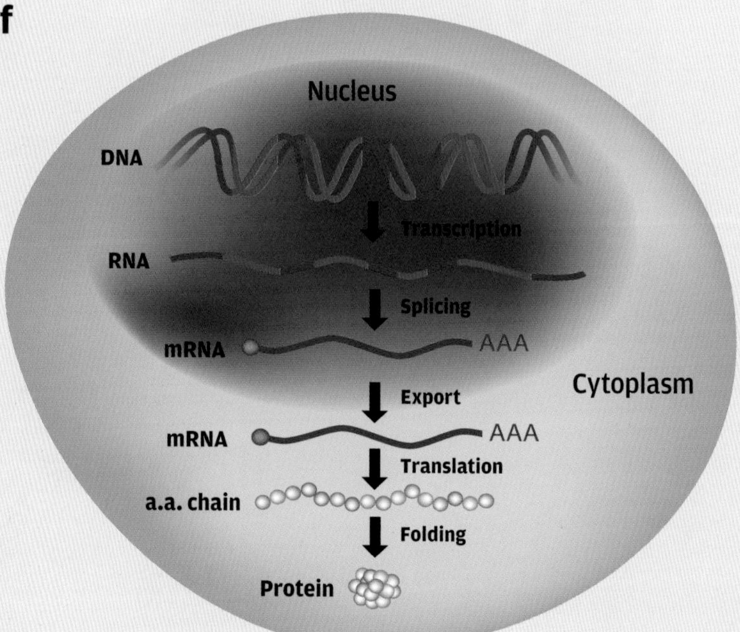

Key processes in gene expression

During the three steps that produce proteins, DNA is first transcribed into messenger RNA (mRNA). The mRNA is then threaded through the ribosomes, which translate it into the protein by assembling a chain of amino acids according to the genetic code.

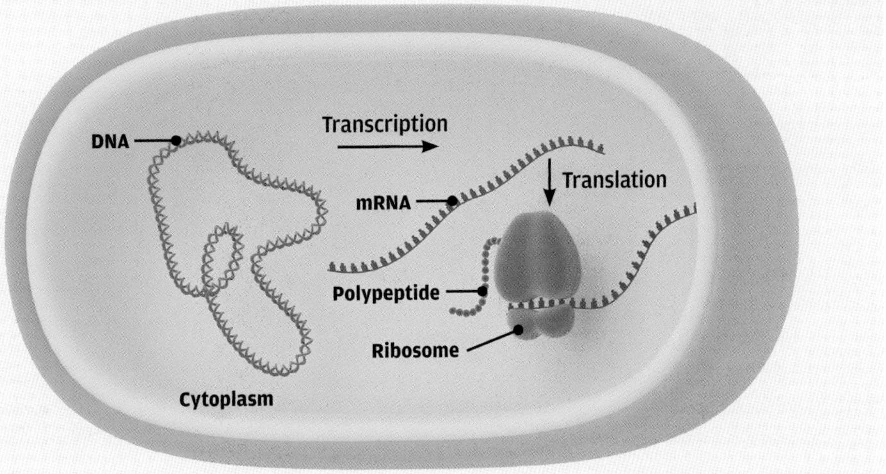

Ribosomes and rRNA

The mRNA is fed through the ribosomes like a strip of film running through a cinema projector. The ribosomes are tiny factories, ready to receive the instructions to make proteins in the nucleus, in the endoplasmic reticulum and floating freely inside the cytoplasm. Partly made up themselves of another kind of RNA, called rRNA, ribosomes not only hold the mRNA in place, so that its code can be read in the correct order, but also introduce the mRNA to single amino-acid-carrying tRNA. Now, at last, the genetic blueprint can be transferred from DNA into protein.

tRNA

This type of RNA acts an agent, matching the mRNA sequence with the amino acid it codes for. This matching process works using a run of three nucleotide bases called the codons. The mRNA sequence is designed so it is read three bases at a time. tRNA molecules are made up of three nucleotide bases called anticodons, because their sequence is complementary to the mRNA codons. Each codon corresponds to a specific amino acid. When mRNA is fed through the ribosomal machinery, the codons are matched with the correct tRNA anticodon, and its amino acid is plucked off and added to the growing chain.

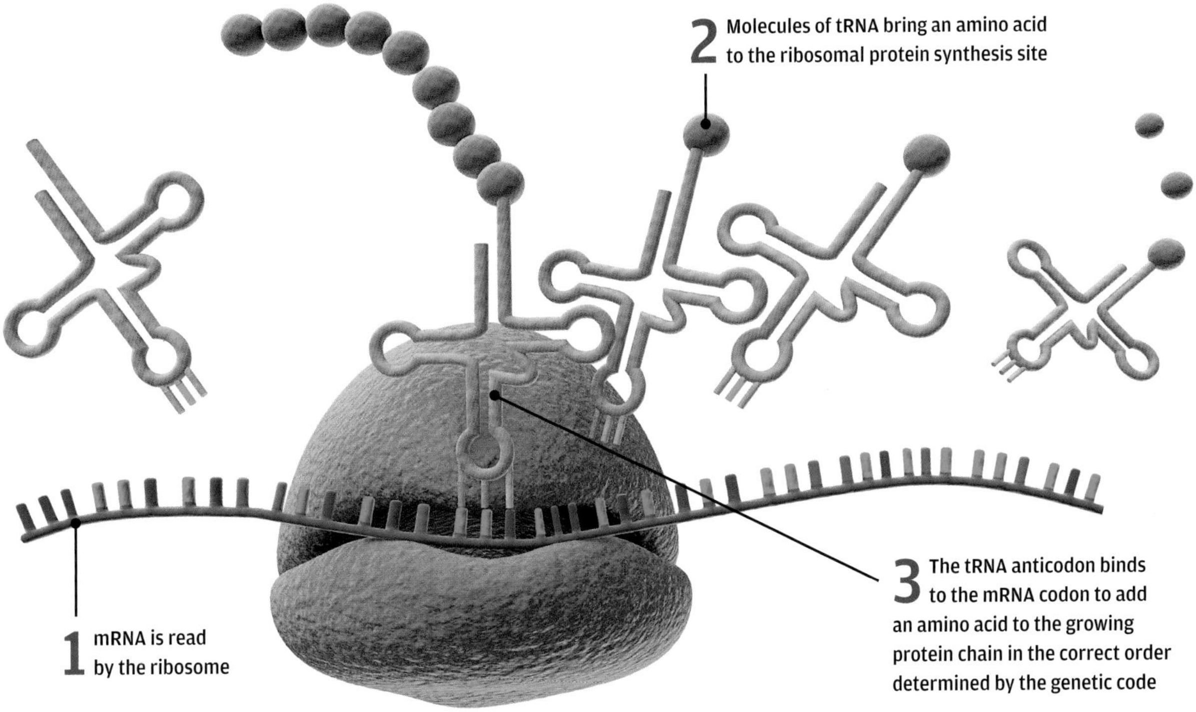

Proteins

Once DNA is transcribed and translated into a growing amino acid chain, proteins really start to take on the form required for them to fulfill their function. But they need to undergo additional changes and modifications to grow into the mature, operational protein structure

Amino acids

The 20 amino acids that make up proteins have different characteristics, which is why one might be chosen over another at a particular position on the protein chain. Some are small, like glycine (Gly), and some are larger, such as tryptophan (Trp). Some have acidic properties and others are neutral. Some are more effective at forming certain shapes, like a triple helix or a pleated sheet, giving the protein and even the body's tissues important functional properties.

Post-translational modifications

A growing amino acid chain is known as a peptide. It becomes longer until it is a fully formed protein and then undergoes a series of modifications to make it fit for purpose. These bespoke adjustments, such as to its size and shape or the addition of chemical compounds or side chains, depend on the intended function of the protein, which will often have to work together with other proteins as a complex. In other words, the proteins have to fit together, and their conformation is key to their role in the human body. It is not currently known how many proteins exist in the human body – scientists are still assembling the proteome, the complete catalogue. However, estimates suggest there are in the region of 70,000 to 90,000 proteins altogether – all made from the genetic code of four bases and 20 amino acids. The very many different types, then, are thanks in part to post-translational modifications.

Haemoglobin

Haemoglobin is an example of a protein designed for the job in hand; transporting oxygen around the body. It is made up of two alpha and two beta chains, assembled in such a way so as to hold iron-containing haem rings in place in a formation designed to carry oxygen.

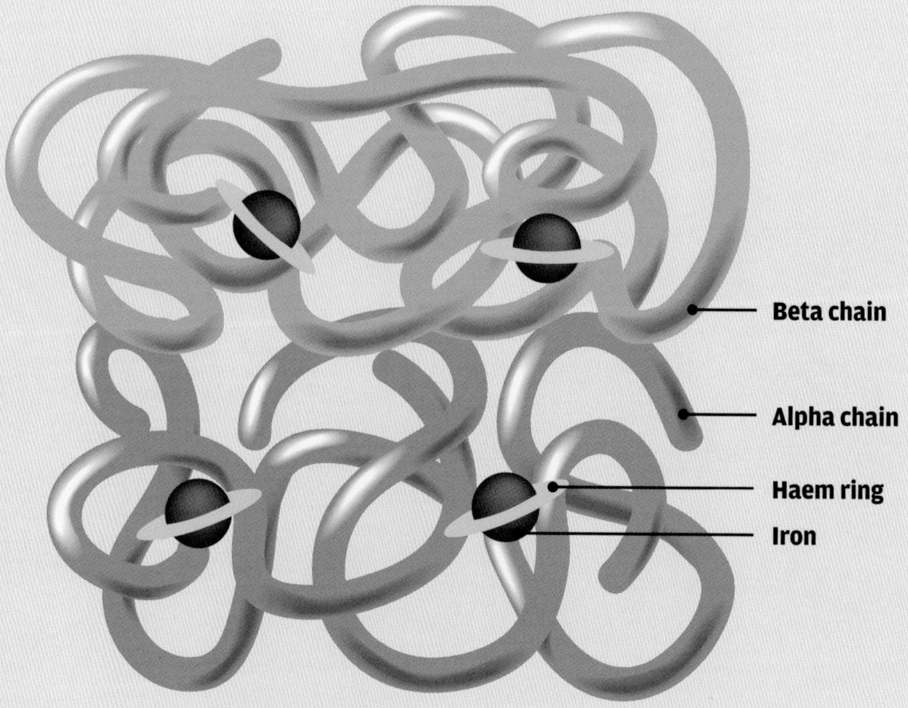

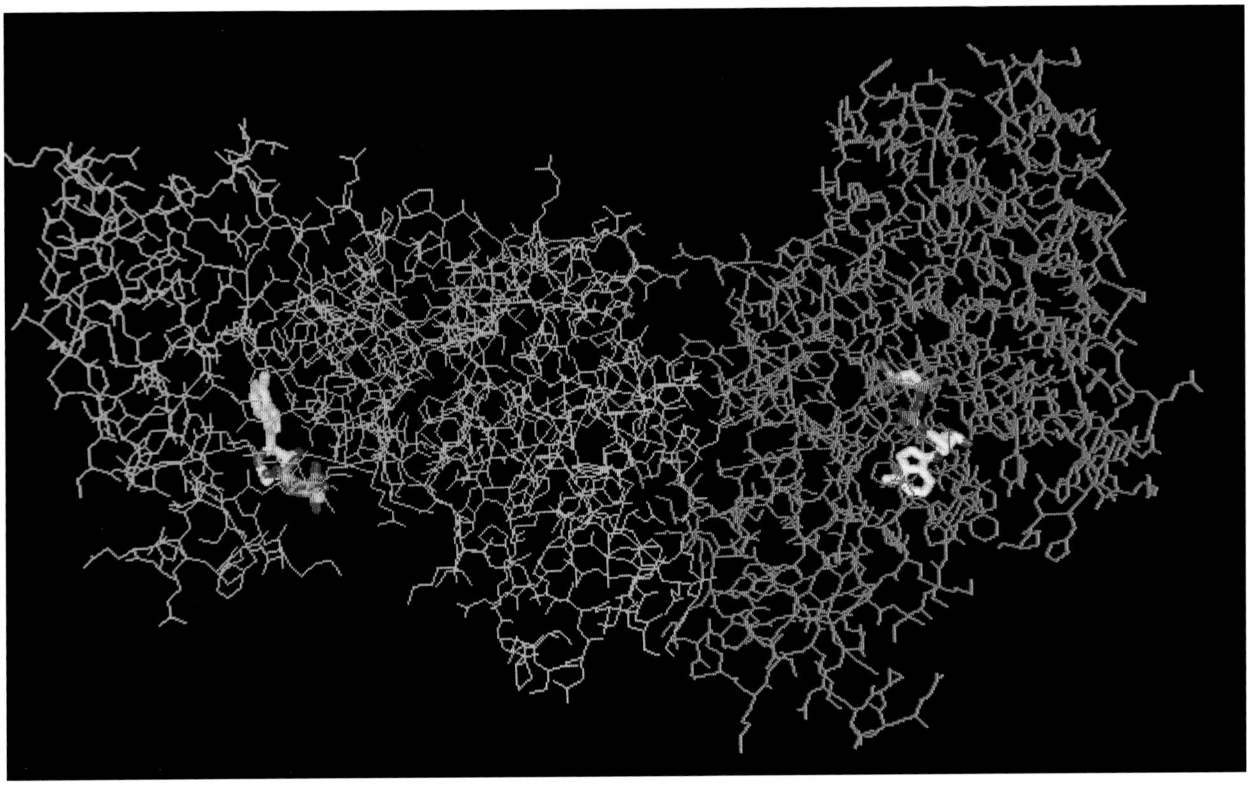

▲ As proteins are constructed, they become increasingly more complex, starting with a polypeptide chain of linked amino acids, which then folds to take on the shape it needs to be to work.

How protein structure works

A good way to understand protein structure is to look at haemoglobin (Hb), which is found in red blood cells. The protein make-up of haemoglobin is designed to transport oxygen from the lungs around the body. Haemoglobin is actually a complex of proteins and consists of four parts, or sub-units, called globin chains. These chains, alpha and beta, form a three-dimensional structure to hold in place a chemical ring called a haem, the part of the haemoglobin molecule that actually carries oxygen. Protein construction is so complex that it takes several genes to build perfectly this essential constituent of the body. Any change to the design could prevent oxygen from binding, because the haem ring could become buried deep inside the protein and is no longer accessible. This is what happens when changes occur in the genetic code.

Exons and introns

Within a gene, only some sections of the DNA encode proteins. These are called the exons. The DNA segments that separate the exons are non-coding and known as introns. Collectively, all the coding exons in the genetic code constitute the exome. Introns actually contain some important sequences that, for example, regulate gene expression or ensure that the non-coding segments are spliced out of the mature mRNA molecule.

'Junk' DNA

Only 2% of the genetic code contains DNA that codes for protein production. Until relatively recently, the remaining 98% was deemed a kind of genetic hinterland, known as 'junk' DNA. Nevertheless, the cell machinery faithfully replicates this DNA each time a cell divides, just as it does the coding regions. But, why such attention to detail if it is supposedly all junk? Because, of course, it is not!

Contained within these non-coding regions are regulatory elements, microsatellite markers and polymorphic variants, microRNAs and transposons, with functional roles in controlling the expression of the genes found in the other 2%. The junk DNA continues to offer up its secrets, revealing fascinating insights into, for example, evolutionary processes and selective advantage. The story will continue to unfold and promises to be spell-binding.

Epigenetics

Not all chemical modifications that occur in our DNA are inheritable. This is the science of epigenetics, a rapidly expanding field of genetics that could explain, for example, why identical twins are not actually identical, why smoking causes cancer and how nature and nurture really do rub along. One of the first-known chemical modifications of DNA that generates an epigenetic effect is methylation. This and other epigenetic changes are explained below.

Methylation

The addition of methyl groups, chemical side chains containing carbon and hydrogen, is really common across the genetic code. Called methylation, this particular modification has a well-defined role in the normal function of DNA, including silencing genes, such as ones that undergo genetic imprinting, but it has also been linked with DNA damage through environmental exposure to toxins, such as cigarette smoking, and to poor nutritional states.

Imprinting

A few hundred of our genes are under the influence of a particular type of epigenetic phenomenon called imprinting. In this is the parent-of-origin effect, where the copy of an imprinted gene inherited from the mother is switched off or silenced through DNA methylation, but the paternal copy remains unmodified and unmethylated, staying switched on. Imprinted genes are often important in the fundamental processes involved in growth and have key roles very early in life. For example, some imprinted genes can influence the size of the placenta or the developing foetus, depending on which parent originated the active form. This creates a situation where two developmental requirements could potentially be in competition with each other.

When these DNA imprinting switches go wrong, significant medical conditions can arise including, for example, the restricted growth condition Russell-Silver syndrome – and its opposite, Beckwith-Wiedemann syndrome.

DNA and epigenetic factors

DNA is under the influence of chemical markers that affect gene expression in a process called methylation. The way it is wound up and packaged also plays a role.

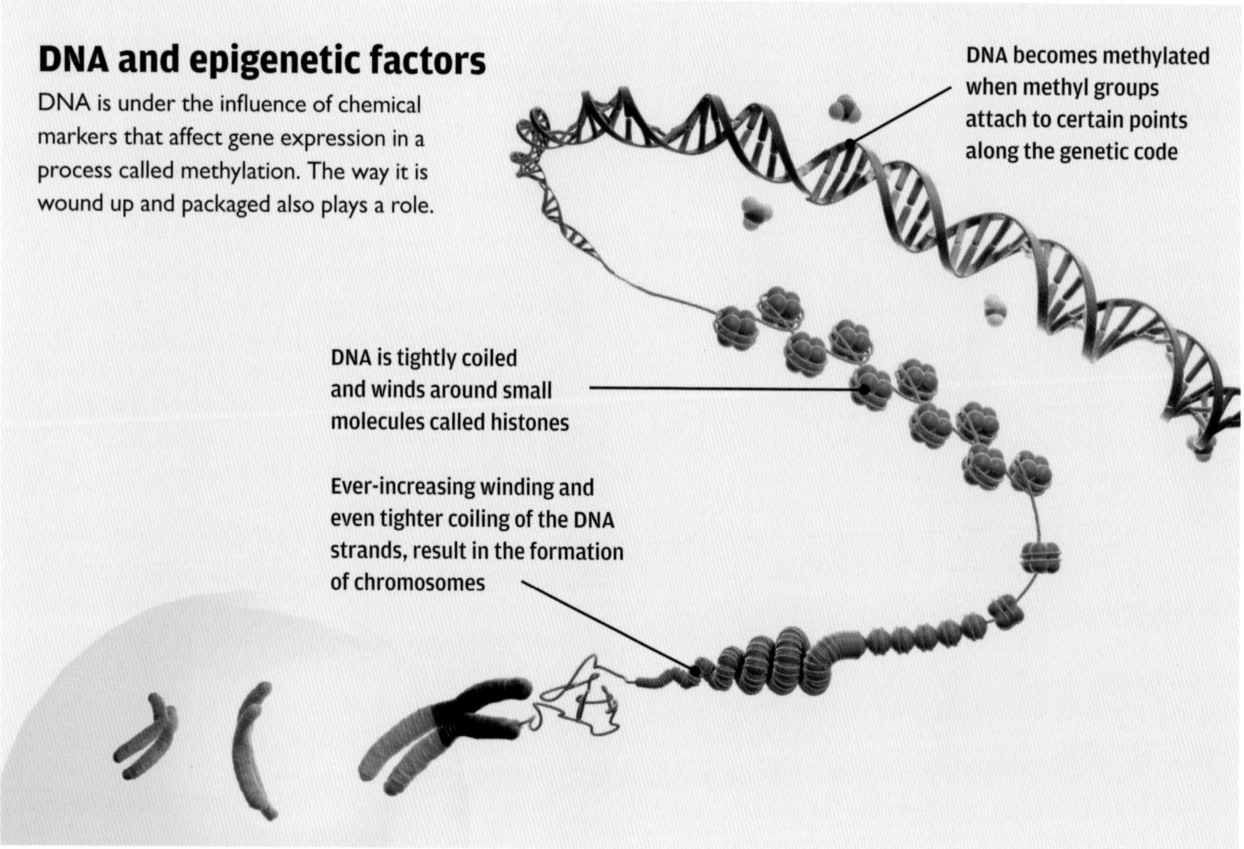

Uniparental disomy

Uniparental disomy (UPD) occurs when both chromosomes originate from one parent. In the case of chromosome 15, if both copies come from the mother and none from the father, a condition called Prader-Willi syndrome occurs. If both originate from the father and no copies of chromosome 15 come from the mother, Angelmann syndrome results.

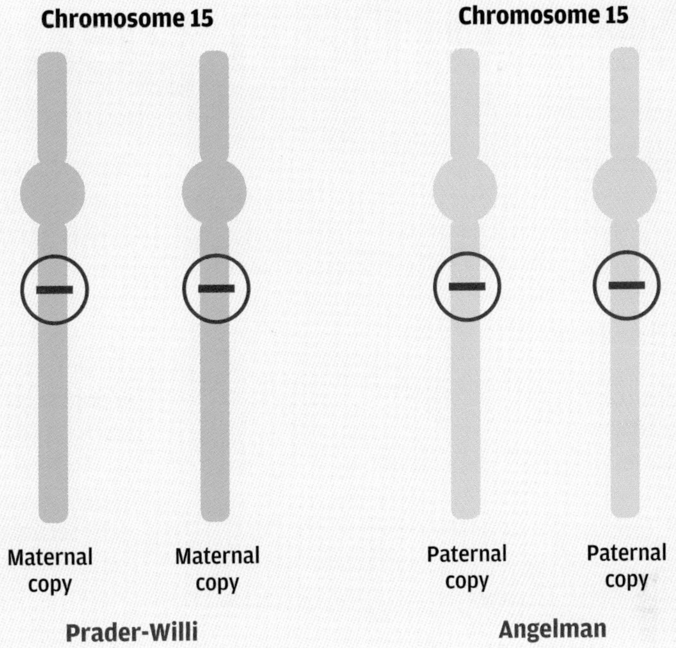

Uniparental disomy

One cause of imprinting disorders is that a baby inherits both chromosomes of a pair from only one parent. This usually arises after trisomy and the cells in the early embryo effectively eject one chromosome in a move to save the situation. This safety net is called trisomy rescue. Maternal uniparental disomy (UPD) occurs when both chromosomes originate from the mother; paternal UPD occurs when they both come from the father. UPD is a problem only if the relevant chromosomes contain imprinted genes that must have representation from both the mother and the father. There are several examples of genetic syndromes occurring through this mechanism.

Prader-Willi syndrome is one such example, and is caused by maternal UPD chromosome 15. People affected with this syndrome want to eat constantly, lack satiety, have intellectual disability, and have slightly unusual facial features called dysmorphism. If UPD 15 is paternal in origin, a different condition arises, Angelman syndrome. Children with this condition have extremely slow head growth, resulting in a microcephaly, severe intellectual disability (never learning to speak, for example) and epileptic seizures. Although the two conditions result from the same genetic mechanism, the effects are completely different, highlighting in extreme circumstances the importance of the parent-of-origin effect. They both affect roughly 1 in 20,000 people.

▼ The DNA is wound around the histone proteins, and methyl groups attach directly to the DNA.

Histone modification

Histones, proteins around which DNA is tightly coiled, ultimately forming chromosomes, are also subject to chemical modification. Adding a methyl or a phosphate group, an acetyl molecule or a ubiquitin protein results in methylation, phosphorylation, acetylation and ubiquitination, respectively. These external chemical tags can influence whether or not genes are expressed and can alter the shape of the coiled DNA structure, again affecting the way the genetic code works.

The full effect of these modifications, in both health and in disease, is only just beginning to come to light. In the battle of nature versus nurture, we still have much to discover.

The human genome

If genetics is the study of heritability, genomics is the study of the complete genetic code – or the genome. This includes all the coding regions of the genes and all the non-coding, so-called junk DNA. To read the genetic code in its entirety requires a number of different methods, including whole genome sequencing, which is finally starting to unlock the secrets held within.

Sequencing

Reading the sequence of a single gene, multiple genes, long stretches of DNA or the entire genome uses a process called sequencing. The way genes are sequenced has changed over the years, starting from the early DNA era to the recent genomics revolution. There are many reasons to study the sequence of part or all of the genetic code, from understanding biological processes to making advances in modern medicine.

Sanger sequencing

A method used to study a relatively short segment of DNA, such as a single exon of a gene, Sanger sequencing is named after Frederick Sanger (1918–2013), an English biochemist and two-time Nobel prize winner. He calculated the bases found in nucleic acid.

Sanger sequencing works out the order of the bases in a DNA strand, calling out the sequence base by base. To do this, scientists must first amplify the stretch of DNA they want to copy (the template), increasing its concentration many thousands of times over. Scientists do this through the polymerase chain reaction (PCR), a three-step cycle, repeated many times. It involves heating the DNA sample to separate the double strands (denaturing), attaching to either end primers, short single strands of DNA whose DNA sequence is complementary to the template (annealing), and using a

Southern blotting

Southern blotting is a method of detecting whole fragments of DNA and calculating their size, which can give an idea of their function.

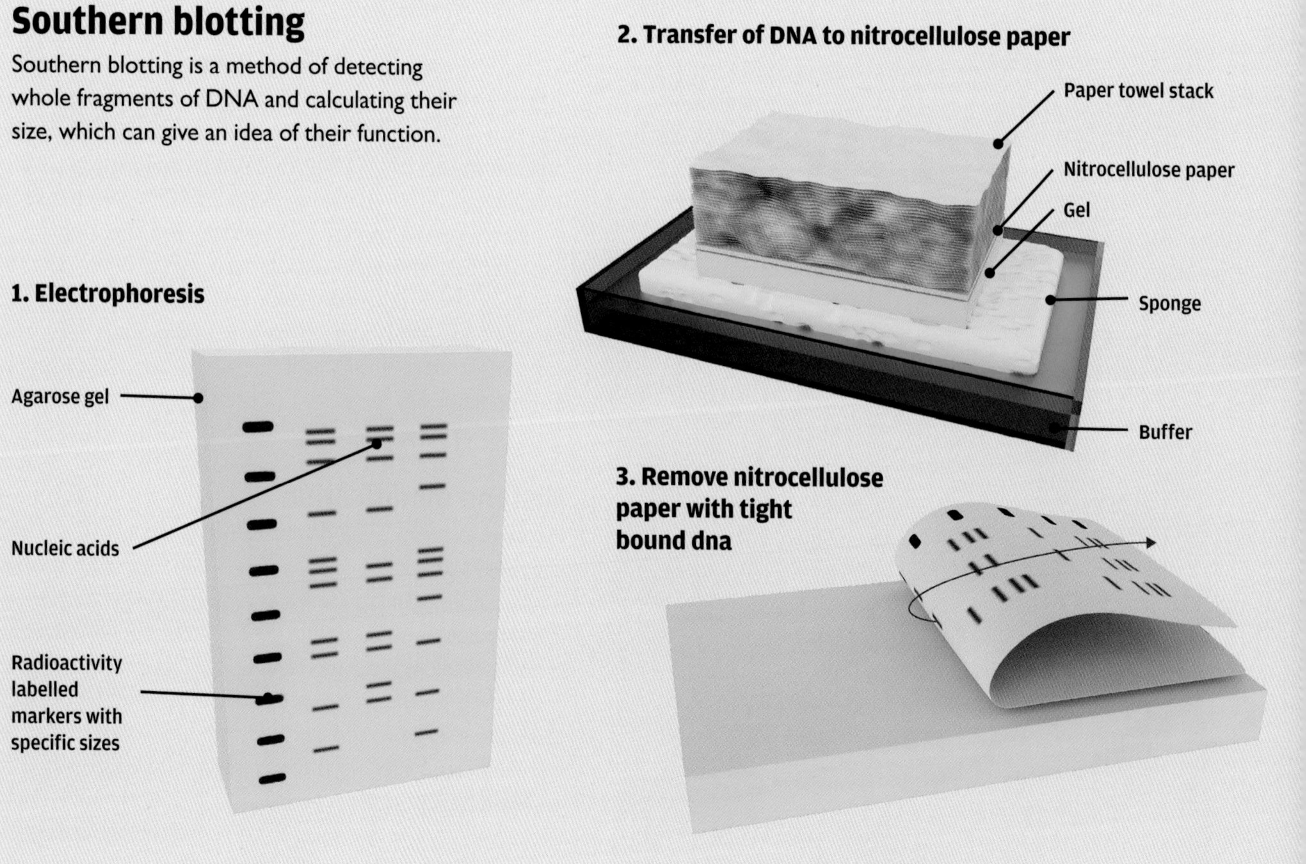

chemical enzyme called polymerase to fill in the gaps between the two primers, creating several thousand copies of the DNA strand of interest (extension).

For the sequencing part, scientists use the PCR process again, but this time using a mixture of regular nucleotide bases with ones that, when they are incorporated at the extension stage, stop that chain from extending any further. This process gives this stage in sequencing its alternative name: the chain termination method. The chain-terminating bases are labelled with differently coloured dyes to indicate whether they are an A, T, C or G in the chain. Researchers can analyse the resulting manufactured strands of DNA using a chromatogram, a system of separating out the bases according to their colour.

Southern blotting

This method of analysis enables researchers to study larger fragments of DNA. It is a way to detect a whole section of DNA, rather than sequencing the individual bases. The first step in doing this is digesting the DNA, chopping it up into smaller fragments using an enzyme (a catalyst protein that triggers a chemical reaction) to recognise a specific

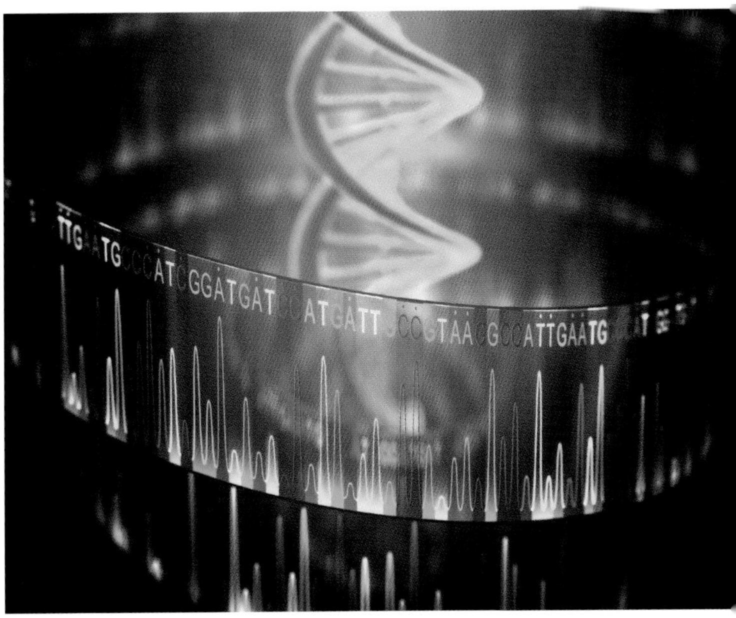

▲ When the DNA is sequenced the genetic code is read as a series of letters that represent the different bases A,T,C&G.

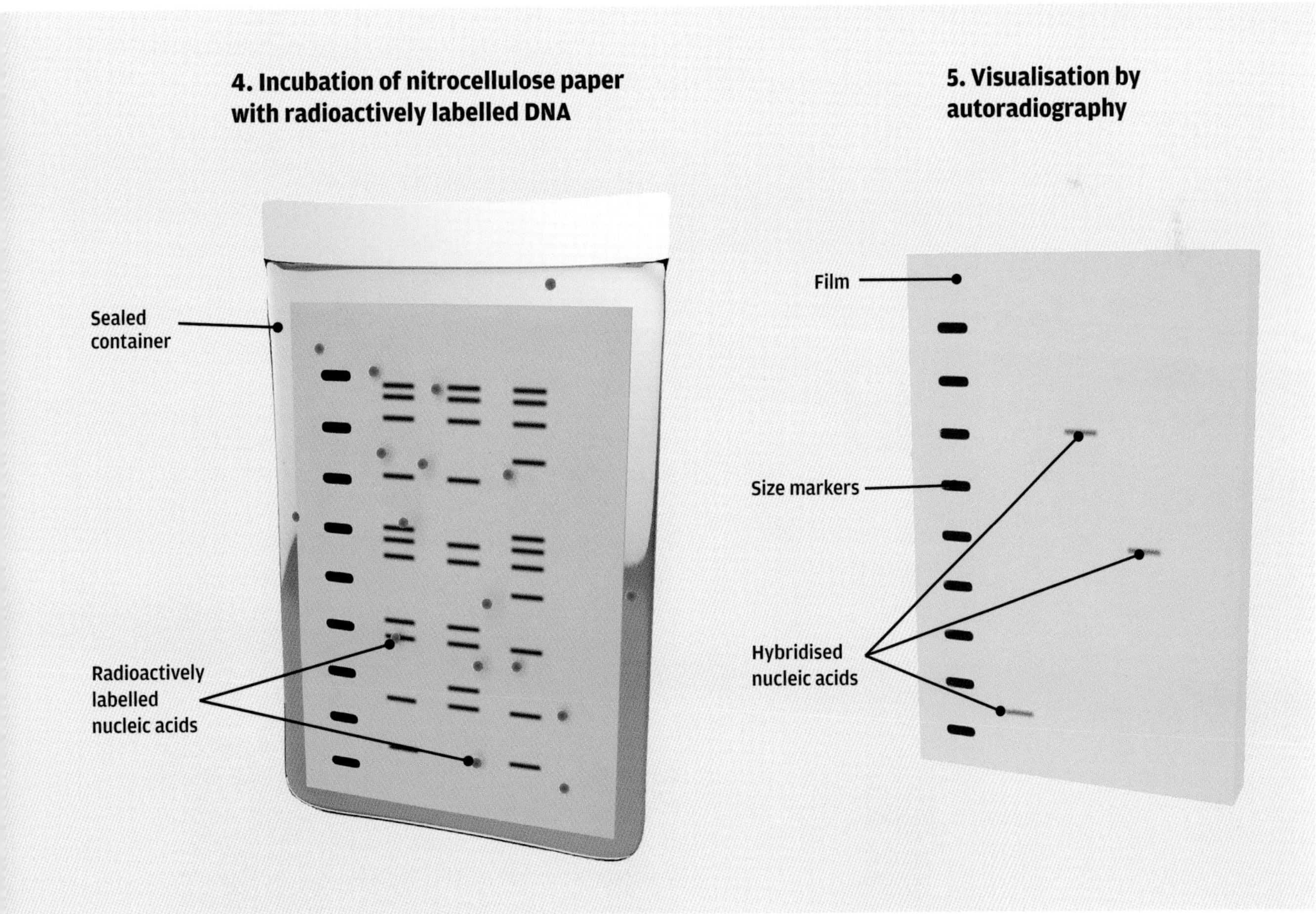

DNA sequence and cut it. The original source of DNA-digesting proteins for Southern blotting was bacteria. Next, the fragments of DNA are sorted into different sizes using a process called electrophoresis. This involves passing an electric current through a special gel, known as agarose gel, to separate them out. The DNA is chemically denatured to make it single stranded, before blotting the fragments on special paper to draw them out of the gel. Complementary strands of DNA, called probes, labelled with a chemical or radioactive tag to help researchers detect them, are applied to the blot to visualise the fragments and determine their size.

Next generation sequencing

Major technological advances in our ability to sequence genes has now brought us firmly into the genomics era. Rather than sequencing tiny fragments of DNA using the labour-intensive Sanger method, scientists can use much higher throughput methods. This revolution in DNA sequencing now means we can sequence the whole genome in one go. Although the technique is expensive, the cost continues to fall.

The process involves the same principles as Sanger sequencing. The first step is to break up the genetic code at random into smaller fragments, which are then assembled in a kind of library. The individual fragments are amplified, just like before, and sequenced, being separated out in an electrophoretic gel. The difference with this method is that thousands of reactions can happen at the same time in what is referred to as 'massively paralleled'.

The result is that many genes, all 20,000 of them, if you want, can be sequenced all in one go. Whole exome sequencing (WES), sequencing those coding exons, and even whole genome sequencing (WGS) are becoming routine practices in genetic analysis. The next challenge, though, is how to interpret the millions, even billions, of bases this high-throughput method produces, and where to store the sequence information generated. Hail! The dawn of bioinformatics and big data!

The Human Genome Project

The Human Genome Project enabled the human genetic code to be completely sequenced from start to finish in an international collaborative effort, lead by US geneticist Francis Collins (1950–), who himself took over the task from James Watson.

In 2003, to great fanfare (and rightly so), the team of geneticists announced the completion of the project. We now know not only that we have 20,000 genes but also how we can map them across the genome. (The number came as a surprise to some, who thought we would have more, especially knowing that a banana has 36,000!) The project

Next generation seqencing

Unlike Sanger sequencing, next generation sequencing, also known as massively parallel sequencing, allows millions of fragments to be sequenced simultaneously.

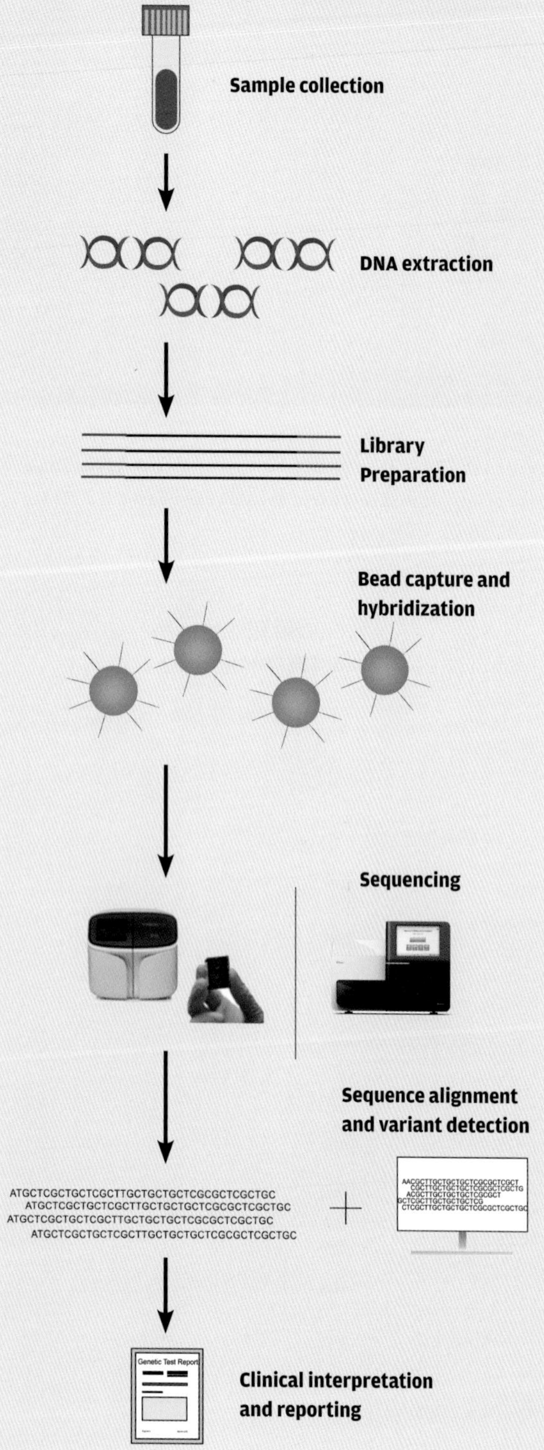

has brought untold benefits to biological science, molecular medicine and evolutionary genetics. It has allowed scientists to create a reference sequence of the human genome, ensuring consistency from one laboratory to the next. It has revealed new information about the 'junk' DNA, which seems to be much more dynamic and functionally important than we thought when it was first given its name.

The 100,000 Genomes Project

Launched in 2015, by December 2018 this government-backed project had sequenced 100,000 genomes, focusing on 85,000 participants with rare genetic diseases and certain types of cancer. The project sequenced DNA from cancer patients' blood and tumours because cancer is thought to disrupt genomic DNA and to drive the personalised medicine agenda in treating different tumours. In including rare diseases, which collectively are not so rare – affecting 1 in 17 people – scientists aimed to end the diagnostic odyssey faced by many affected children, explaining the nature of their condition by harnessing the information held within their genomes.

The project promises to transform genomic testing in modern-day clinical practice, making such testing a mainstay of medical treatment and helping to develop bespoke therapeutics based primarily on the profile of an individual's genetic code, rather than using a one-size-fits-all approach to medical intervention.

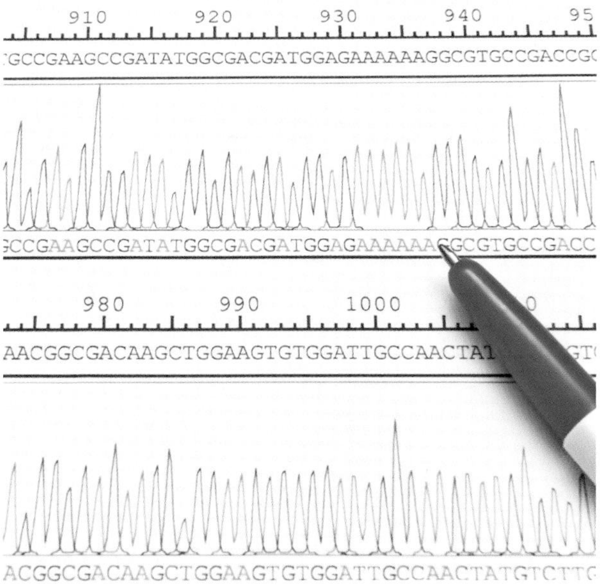

▲ Sanger sequencing produces a read out of the DNA called a chromatogram, a trace that is created by differently coloured labels specific to each of the bases A,C,T and G. It represents the final read out of the DNA sequence.

▼ A technician loading samples into automated DNA sequencers, used for analysis carried out by the Human Genome Project.

CHAPTER 2

Genetics in action

People talk about DNA being the key that unlocks the code of life. That code manifests itself in the intricate make-up of proteins, themselves the combination of amino acids, that together give function to all the cells in our body. This chapter explores the nature of the genetic code and the way in which it enables the body to function healthily and gives us each our uniqueness.

What is the genetic code?

Today, when we think about what a code is, many different things come to mind, depending on the context. A code can be a system of rules that we must read in a certain way, so that we can convert the information held within the code into something else. Code can be something hidden in a message as a way to convey something in secret to another person. Morse code, a system of communication that uses a series of dots and dashes to represent different letters and numbers, was originally developed to send telegraph messages over long distances. We apply the term 'code' to software programming, 'coding' a reference to the step-by-step instructions that a computer needs to complete a required task. An organisation's code of practice is a set of regulations its members must follow – how they should behave and how they should dress, for example.

In the case of the genetic code, all of these elements apply. It was secret until, in 1961, Francis Crick decoded the hidden messages held within it. The genetic code forms the step-by-step instructions living organisms need in order to be built. It is comprised of different 'symbols' – the four nucleotides, A, T, C and G – needed to confer long-distance messages across 3 billion base pairs. And it abides by strict regulations, a cellular modus operandi that can have devastating effects in the form of genetic defects if it is not followed to the letter. Simply put, the genetic code is the set of instructions contained within DNA that all living things need to construct the building blocks of life – proteins.

▼ Proteins like this one, NOTCH1, have a structure that lets them carry out a certain function on the body.

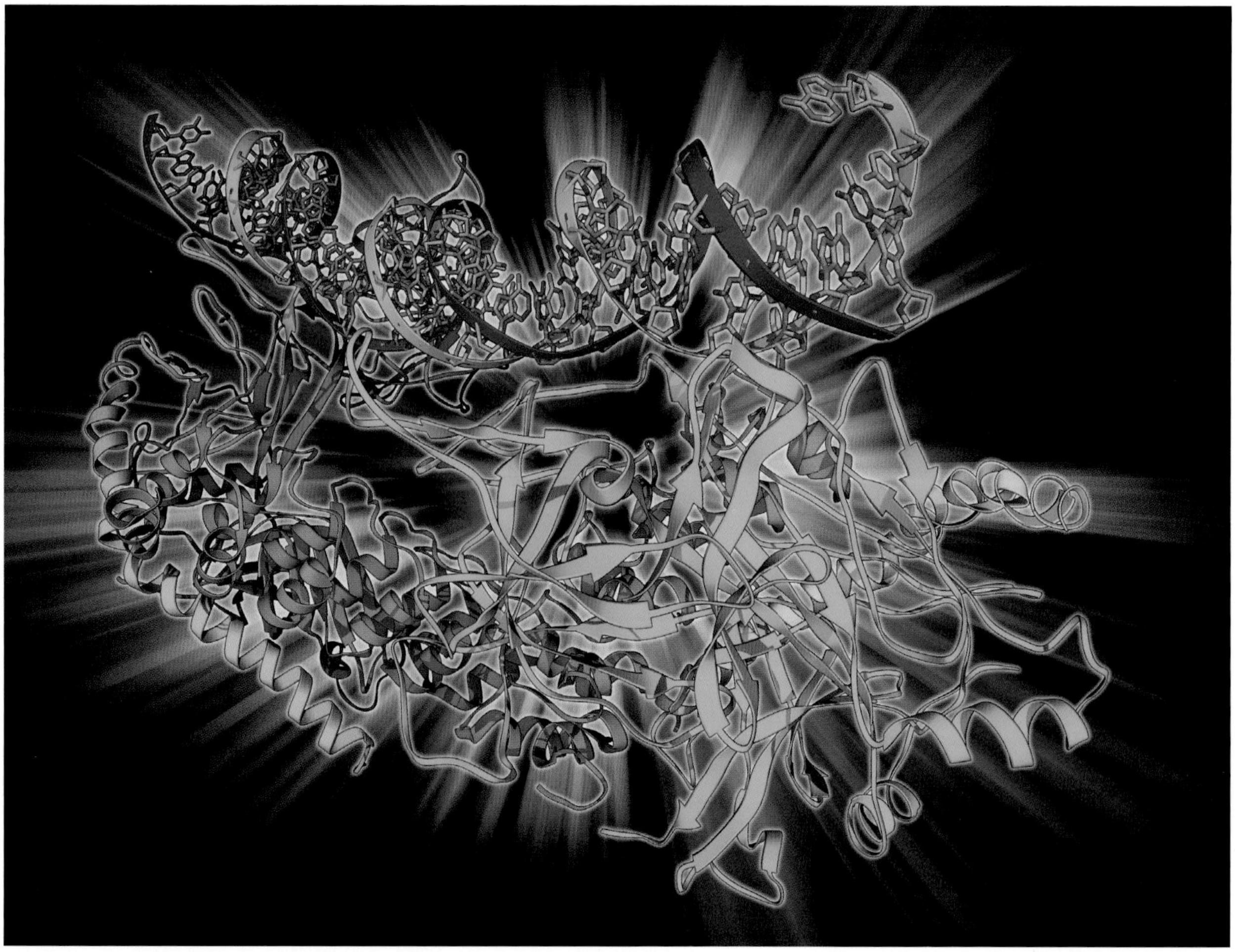

Red blood cells

Red blood cells (right) contain the protein haemoglobin (left), both of which are designed to transport oxygen around the body.

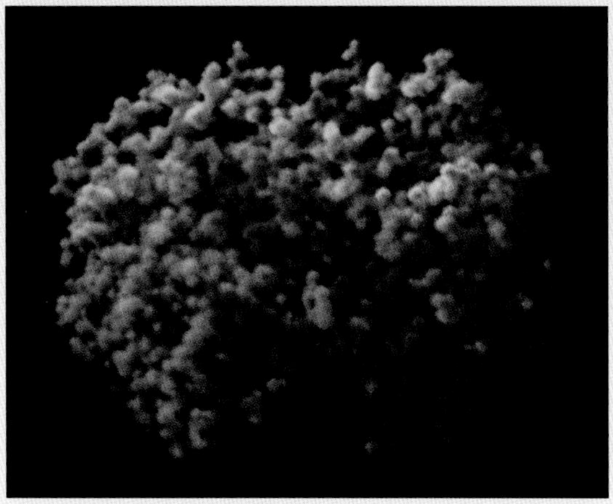

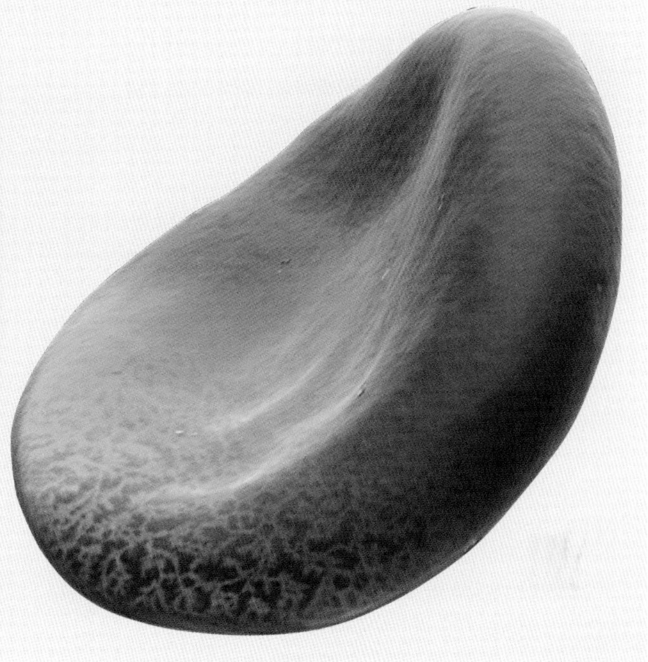

DNA: protein by protein

The genetic code is the blueprint for life. Fundamentally, it is a collection of DNA recipes for the thousands of proteins that make up the 200 different tissues in the human body.

Proteins are the building blocks required for almost every function of our cells. They protect us from infection in the form of antibodies. We can walk around because the proteins in our muscles have the ability to contract, giving our bones and joints movement. They are the stuff of enzymes, which control a multitude of chemical reactions in our body. Hormones, made up of protein derivatives, send messages to our organs, controlling growth and development, making us feel hungry or full, regulating our metabolism and responding to physical stress. The formation of proteins gives our organ tissues their ability to function according to their specific design. For example, it's because of the specific proteins in our liver cells that the liver has the power of detoxification, and, in our kidney cells, that the kidneys have the capacity to excrete toxins. The ability of some specialist cells to store macromolecules, such as fat, means we have an insulating layer of adipose tissue under the skin, so we can store energy until we need it. Some proteins – such as haemoglobin – can transport other chemicals and compounds around the body.

Haemoglobin: the oxygen transporter

The function of haemoglobin is to transport oxygen from the lungs around the body in the red blood cells. Haemoglobin is a globular protein, built with stepwise complexity from its basic amino-acid chain to the finished product, a compact sphere, or globe, that sits inside the red blood cell. It contains four separate proteins called sub-units, two alpha-globin chains and two beta-globin chains. These link together, creating a sphere with a space in the centre inside which sits the iron-containing haem ring, a protein facilitator or co-factor that holds the oxygen molecule.

Protein structure

Building specialised proteins requires four different levels of structure, with increasing complexity across the different manufacturing stages. The primary level is the order of amino acids, translated directly from the genetic code into the polypeptide chain. This chain then coils and folds in on itself, as the different component amino acids interact with each other, bonding to create a three-dimensional, secondary level protein structure. Further adaptations to make sure it is held strongly in place occur at the next stage, the tertiary level, so that it can hold its shape and form the quaternary level structure that brings together the separate alpha and beta sub-units along with the cofactor haem ring.

But the starting point is the polypeptide chain, the basic structure needed to create a mature protein molecule, making it fit for purpose. The exact order of the amino acids, joined together by interlinking peptide bonds, is critical to construct a protein of the right size and shape for its intended job in the body.

When haemoglobin goes wrong

Different genes encode the separate sub-units and combine as the fully formed haemoglobin protein. Two genes hold the code for the alpha sub-units 1 and 2 (*HBA1* and *HBA2*) and one gene for the beta sub-unit (*HBB*). Defects that result in damage to the alpha or beta chains cause genetic forms of anaemia, low concentration of haemoglobin in the blood. Known as alpha and beta thalassaemia and sickle cell disease, these conditions are more common in people from certain ethnic backgrounds. For thalassaemia, those originating from areas such as the Mediterranean, southern Asia and the Middle East are more susceptible; and for sickle cell disease, those originating from Africa are more susceptible. People with these conditions must inherit a fault in the same gene from both parents, as they are all autosomal recessive disorders, meaning that the error in the code sits on a recessive gene.

In thalassaemia, defects in one of the three haemoglobin genes leads to reduced amounts of that chain being made. If it is one of the alpha sub-units that is faulty, the affected person will have alpha thalassaemia; if it is the beta sub-unit gene, he or she has beta thalassaemia. There is a range of severity in the thalassaemia conditions, depending on how much haemogolobin the body can produce. A sufferer may need regular blood transfusions to help the body overcome the persistent anaemia.

Sickle cell disease is caused by a fault that affected the haemoglobin beta sub-unit. It is different to beta thalassaemia, because the red blood cells form a crescent or sickle shape, rather than a globe. Affected people have anaemia because their red blood cells do not live as long as regular ones, and they are also at risk of sickle cell crises. These painful and sometimes life-threatening events occur when the abnormally shaped red blood cells become stuck in smaller blood vessels, blocking them and preventing oxygen getting to the tissues. Treatments of sickle cell disease include regular painkillers and blood transfusions, as well as avoiding situations that can precipitate a sickle cell crisis, such as conditions in which oxygen concentration is low, at high altitude, say.

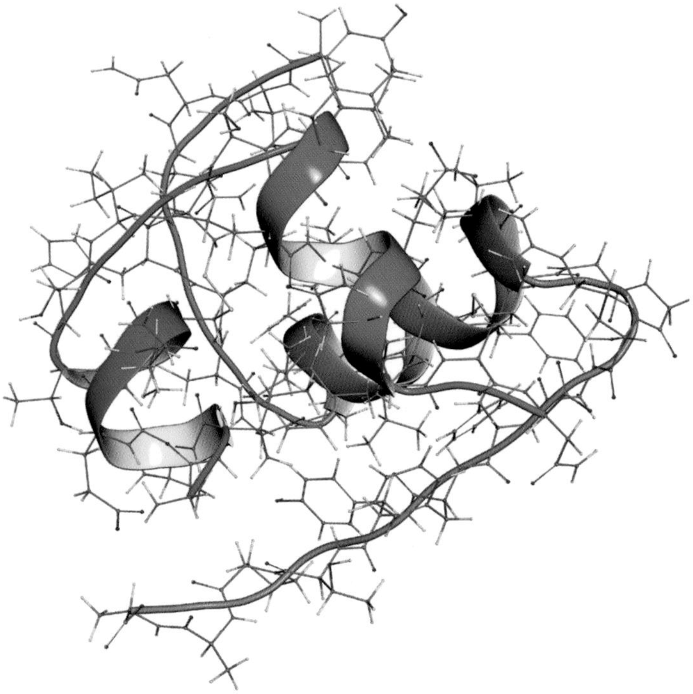

Amino acids

So what is it about amino acids that provides the wherewithal to accomplish this? The body's 20 different amino acids share the same basic structure: an amino group of one nitrogen and two hydrogen atoms (NH2), a single hydrogen atom (H) and a carboxyl group of carbon, oxygen and hydrogen (COOH). The amino acids each have slightly different physical properties because they each have a variable or 'R' group. The distinctions brought about by the R group might relate to individual size, electric charge and ability to link with other amino acids and cofactors. Put an amino acid in the wrong place and the protein could lose its ability to do its intended job efficiently.

Collagen: the master builder

Collagen is the most abundant protein in the human body, making up about one-third of the total proteins we have and being widely distributed. It is found in the connective tissues,

◀ As proteins are created from a growing chain of amino acids called a polypeptide, their structure becomes more and more complex, like here in the case of the hormone, insulin.

which hold our bodies together. It is in our muscles and blood vessels, our bones and cartilage, and our tendons and ligaments. You'll also find it inside the gut, the corneas of our eyes, the kidneys and the skin. There are at least 28 types of collagen, denoted by the Roman numerals I to XXVIII, making up this super-family of proteins, each with different physical properties.

Most types of collagen are described as fibrillar because they are made of supercoiled and entwined fibres known as fibrils. The individual polypeptide chains twist to form this quaternary protein structure, with three strands binding to form a triple helix. The fibrillar collagens type I, II and III make up 80–90% of all collagen proteins. Found in the skin, bones, cartilage and connective tissues, they take up the external space or matrix where the cells key to the workings of these different tissues exist.

Their widespread distribution is clearly illustrated when we consider the effects of genetic diseases that disrupt their production or function. For example, insufficient type-I collagen causes brittle bone disease. Defects in type-II collagen can cause a range of disorders from early onset arthritis to severe dwarfism. Abnormalities of type-III collagen cause Ehlers-Danlos syndrome, a condition causing extreme flexibility of the joints, chronic pain, and stretch marks even if the affected person is slim.

▼ There are 20 amino acids, all sharing the same basic amino structure, but each with a unique variable group to give them different physical properties.

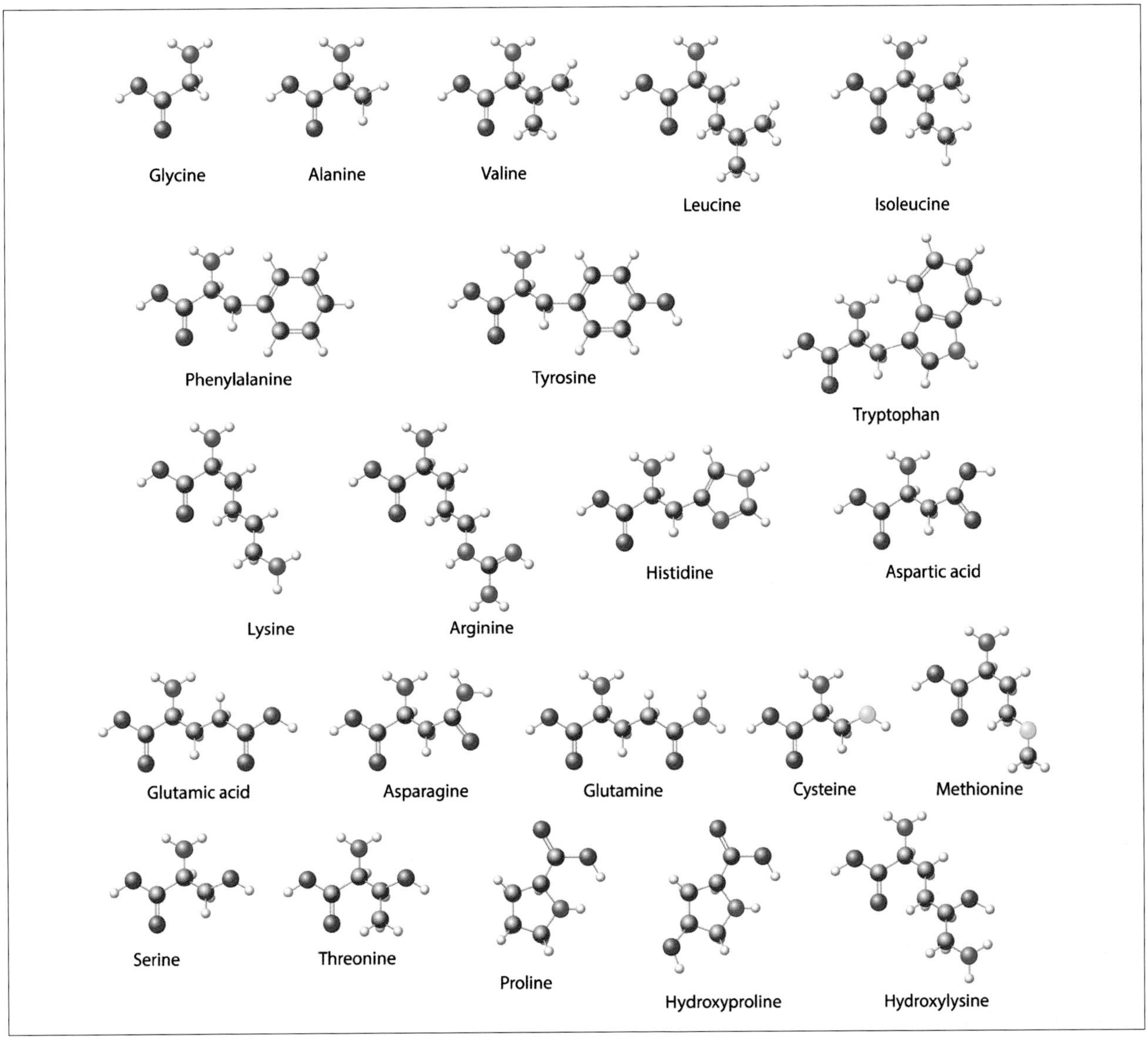

The genetic code recipe book

Genetic conditions such as those mentioned – sickle cell anaemia, Ehlers-Danlos Syndrome, brittle bone disease etc – emphasise how important it is for the body to build the right protein in the right place at the right time, to the letter of the genetic code. How does it do this? It follows a recipe.

We can organise the recipe book for the genetic code into different sections, just like any comprehensive cook book. But instead of starters, main courses and desserts, we find the categories enzymes, structural proteins, hormones, transporters, immunity and so on. It is a big book – there are 20,000 recipes. They might not be neatly divided into individual volumes – so that, for example, the instructions for antibodies are listed on chromosome one and the special kidney edition is on chromosome two – but they are actually arranged in a seemingly random order across the whole genetic code.

Which recipe is required and when is part of the magic held within the code. There are also genes encoding proteins called transcription factors that attach to the DNA at the start of its recipe, instructing it to 'start cooking'. This initiation command means that genes are activated to make their protein only when and where they should. In practice, the construction of one protein triggers a protein-manufacturing chain reaction in a series of interlinked events, coinciding to form vast networks of proteins and interacting pathways, signalling cascades, interwoven matrices, transport and defence systems, to conduct locomotor activities, to repair damaged tissues, to replicate and to reproduce, the whole nine yards.

▼ Our genetic code contains the complete recipe to make a human being.

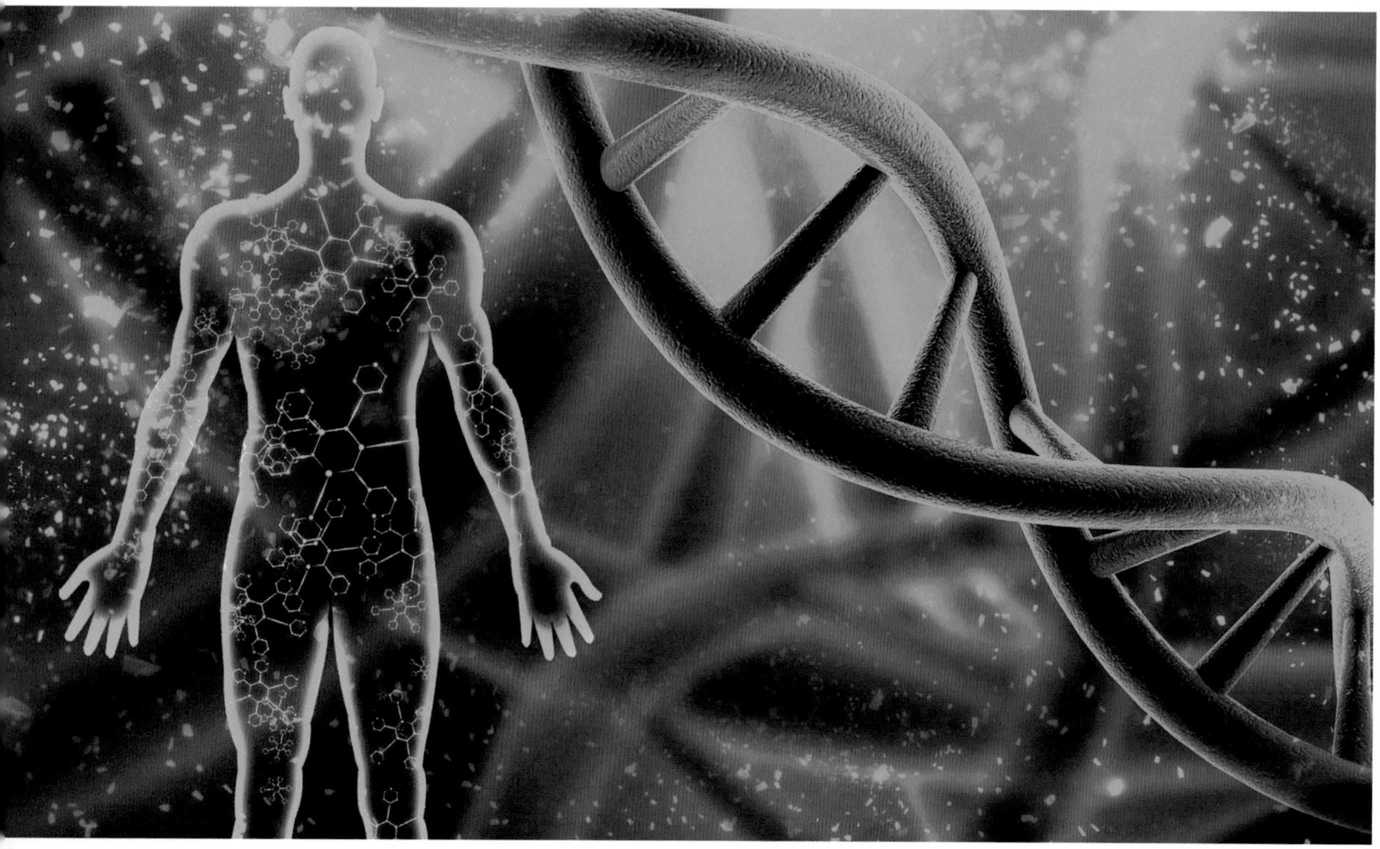

COL1A1 recipe taken from 'The complete genetic code'*

Everyday type-I collagen synthesis

There's nothing like type-I collagen to give your bones the strength they need to get you out and about! Any card-carrying human being should give this recipe a go. Master this and you will be locomotoring!

Difficulty:

intermediate

Preparation:

Transcription – 30 seconds
Translation – 25 seconds
Total cooking time – 55 seconds, plus time for post-translational modifications

Make ahead:

The finished collagen requires two strands of the alpha-1 chain and one of the alpha-2. When everything combines and forms a triple helix, believe me, you'll be jumping for joy! Complete the recipe for the alpha-2 chain in advance and store in your endoplasmic reticulum until needed for the final construction. (NB the genetic code recipe for this, *COL1A2*, can be found in a completely different volume – not on chromosome 7 with *COL1A1*, but on chromosome 17.)

Expert's tip

You can buy pre-prepared type-I collagen from healthfood stores or online, but beware the false claims that it can be absorbed through your skin to reboot elasticity – it's far too large a molecule for that. Better to synthesise your own following this recipe and enhance its natural production with vitamin C.

Use a ready-made tRNA-amino acid mixture – the three-base tRNA codon always links to the same amino acid, just as that masterchef of genetics, Francis Crick, once taught us.

Know-how

Polymerase is an enzyme we need in order to make strands of DNA and RNA. You shouldn't have any difficulty getting hold of some – it is everywhere in your cell store cupboard.

Spliceosomes are cellular machines, programmed to find the start-and-stop sequences inside the RNA.

Ingredients

For transcription step

Transcription factors (we used *RUNX2*, but osterix works just as well)
Ribonucleic acid (RNA) bases A, U, C and G
RNA polymerase
Spliceosomes

For translation step

Ribosomes
Ribosomal RNA (rRNA)
Transfer RNA (tRNA)
Amino acids – all of them, plus extra glycines, prolines and lysines

For post-translational modification

1 alpha 2 collagen 1 chain
Enzymes: hydroxylase and transferase
Vitamin C
Sugars (glucose and galactose work best)
Alpha-2 chains (see make ahead)
Oligosaccharide, such as mannose

To assemble the collagen fibre

Lysyl oxidase
Copper

▼ Collagen fibres are found in many tissues of the body, including bones, tendons and skin. They provide strength in numbers.

* as imagined by the author

Method

1 Make your messenger RNA (mRNA). This is best done in the nucleus of fibroblast cells from inside connective tissue. Try to keep the cells at body temperature throughout. To start gene transcription from the *COL1A1* DNA of your genetic code, add your transcription factor, *RUNX2* or osterix. You should find that it binds easily; if it doesn't, you might be using the wrong ingredients.

2 Allow your *COL1A1* DNA double helix to unzip slightly and add the RNA nucleotide bases, A, U, C and G. Now mix in the RNA polymerase. If your fibroblasts are at the right temperature, then the mRNA chain will start to grow as the RNA molecules line up opposite their complementary DNA counterparts. Within a few seconds you will have your first mRNA chain, a primary transcript called pre-mRNA. But it will need trimming to get rid of the useless junk that is interspersed throughout – this might seem like unnecessary attention to detail, but it is worth the effort!

3 Mix your spliceosomes with the pre-mRNA. Allow them to roll along the molecule until they encounter a 'stop' sign. Snip it here, so the junk impurities can be spliced out. Carry on along the pre-mRNA until your spliceosome comes to a 'start' signal again and snip. Next, join the ends of the mRNA chain together – you should have some polymerase left over from the previous step to do this. Repeat this over and over again until you have your full-length, mature mRNA transcript. If all has gone to plan so far, it should be about 1,500 bases long. Now it is time for the important translation stage.

4 You can do this step in either the nucleus or the endoplasmic reticulum. Why not use both to make as much collagen as possible. Carefully feed the mRNA transcript through the ribosome-rRNA processor. Add the tRNA-amino-acid mixture (see expert's tip) and watch how the mRNA three-base codon finds its tRNA anticodon opposite. Codon by codon, the transcript is translated into amino acids, the peptide chain growing with invisible chemical bonds linking them together. You should find in the final polypeptide chain, the order goes glycine, then proline,

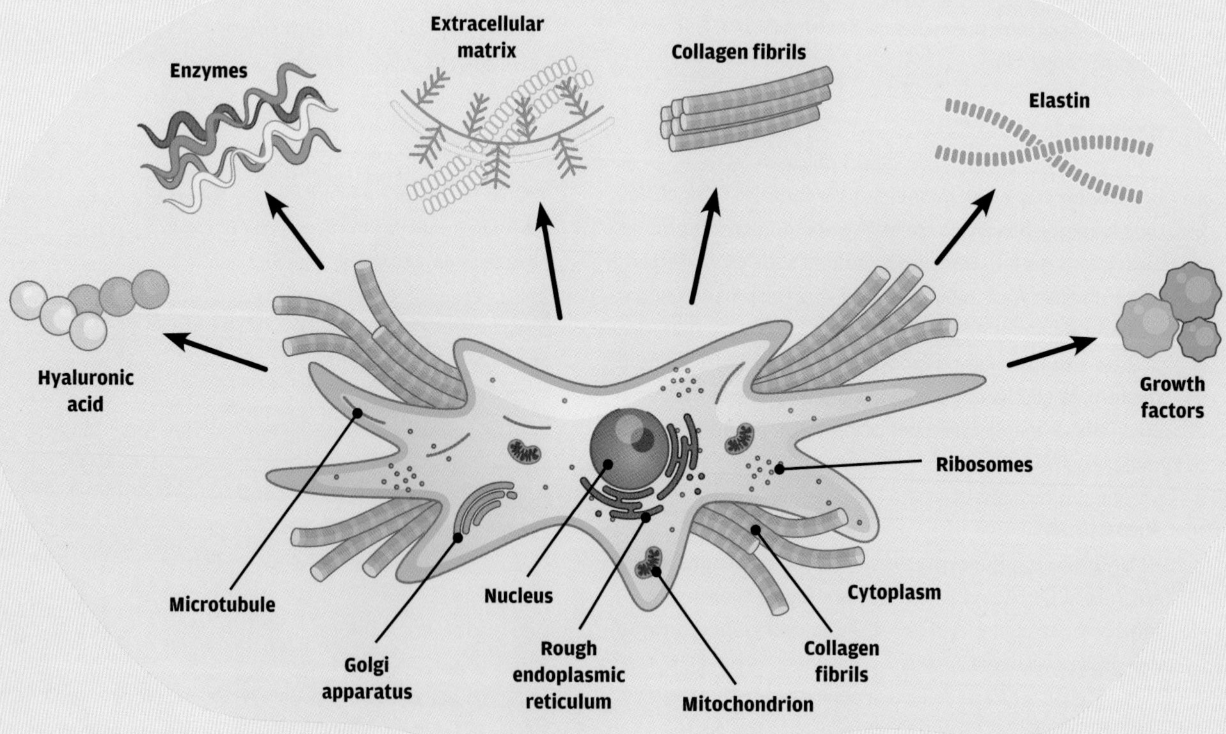

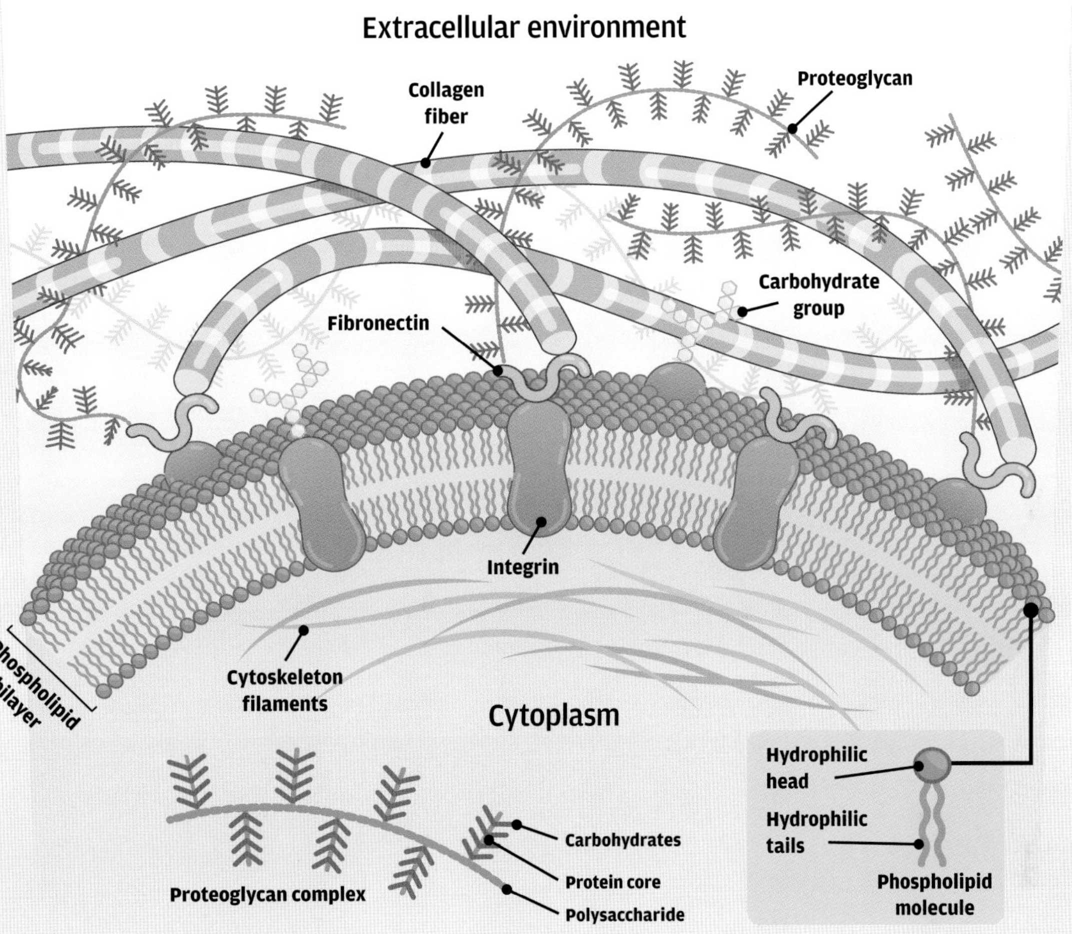

followed by a non-glycine, non-proline amino acid, repeated over and over again until it is 500 amino acids long. This is important for the next step, post-translational modification.

5 It is best to do this step in the endoplasmic reticulum; make sure you have your Golgi apparatus to hand. Trim one end of the polypeptide, then add the hydroxylase enzymes along with the vitamin C. When you do this, lysine and proline amino acids become hydroxylated, which helps the chain make the characteristic crosslinks in the finished product. This key step will not work without vitamin C, so make sure you always have adequate supplies. Next, add the transferase enzyme along with the sugars, so that they attach to the hydroxylated lysine amino acids.

6 Next, working in the Golgi, add the pre-prepared alpha-2 chains to the alpha-1 mixture. Because of the chemical attractions created in the previous stage, two of the alpha-1 chain and one alpha-2 will come together, forming a triple helix. Add the oligosaccharide sugar for the finishing touches. Then the cell will secrete the protein into the matrix space outside.

7 For the final step in synthesising the collagen fibril, add the enzyme lysyl oxidase, making sure that you do not forget the copper – synthesising will not happen without it. The lysines and hydroxylysines can now bond together, linking several triple-helix strands to form the mature type-1 collagen fibril.

Makes enough type-1 collagen to supply connective tissue throughout the bones, joints, tendons, ligaments, skin, corneas and teeth. Goes well with other collagen types, such as II, IX and XI.

From bases to body

We know, then, that sections of DNA within the genetic code are transcribed into RNA and this is translated into proteins. We know that proteins are strings of amino acids that require tweaking and finessing to construct the finished three-dimensional product equipped to do its job. From the moment of our conception, when the egg and sperm unite, the protein-manufacturing machine fires on all cylinders. Every basic biological function from the start of life to the end is reliant upon the finely tuned processes controlling gene expression.

Some of the first proteins to be made are ones that determine which side of the embryo is to be the left and which side the right. This is a process called developmental patterning. Creating a left–right axis is all down to gene expression. The same goes for anterior–posterior positioning that works out front from back; and for cephalo-caudal orientation, which decides top to bottom. The heart orientates with its apex pointing to the left side, and the liver grows on the opposite side. We have four limbs, two upper and two lower, each with five digits, although the fingers and thumbs are on our hands and the toes are on our feet, these being located at the opposite end to where the head is.

There are a handful of genes responsible for male and female development, according to chromosomal make-up – 46,XX or 46,XY. Growth of all the internal organs is determined by sections of the genetic code being activated at the right time, as is the circulation. There are genes that are switched on during prenatal life – for example, the one that makes a form of haemoglobin designed just for foetal life inside the womb. Birth itself is under genetic influences too – another feat of perfect timing, when genes inside the placenta control the expression of proteins that cause contractions, resulting in delivery of the baby.

During the first year of life, a time of rapid growth, a

Of mice and men

We might not look anything like fluffy rodents, but mice and humans have a whole lot in common at the genetic level. Along with many other mammals, we share a common ancestor, dating from some 80 million years ago. Remnants of this common ancestry can still be observed in our DNA today. Our genetic codes are roughly the same size. The human catalogue of proteins, the building blocks that make us, is 85% the same as the mouse. Some genes are 99% identical. These are the ones that are fundamentally important to animal existence. (Studies across many species show that certain regions of the genetic code must remain the same.)

There are significant differences between the genetic codes of mice and humans, of course. We can walk on two legs and mice scamper about on four; they have central incisor teeth that keep on growing, while ours do not; they have a tail and we do not; and of course we have higher cognitive function than mice. Enough similarities remain, though, that we can study mice as a proxy to humans in our understanding of developmental and genetic disorders. Scientific information gathered in this way is now helping researchers to find new treatments for rare diseases – discoveries that are set to transform medicine.

baby can triple in weight and grow to be one-and-a-half times longer, much of this under genetic influences, as well as nutrition. As a baby becomes older, muscles develop some more and bones strengthen and reshape, allowing the infant human body to move around independently. Babies produce new hormones, like growth hormone, and their immune systems make thousands of antibodies to all the 'foreign' proteins their systems encounter, fighting off potential infections. The gut adapts to absorb new foods, and the brain grows and matures. The internal organs function 24 hours a day, seven days a week. Tissues repair and regenerate and replicate, over and over again.

We go through puberty at a genetically determined time, the result of genes producing another set of hormones. At this stage our bodies prepare to create the next generation, through reproduction reliant upon regular sperm production in males and carefully orchestrated meiosis forming mature eggs in females – all under genetic control.

With increasing age, gene expression becomes less efficient and we start to show signs of ageing and of degeneration. The ends of chromosomes, the telomeres, become shorter and the array of epigenetic markers changes. Some cells cannot fend off the impact of environmental insults any longer and let mistakes copying the genetic code at cell division go unchecked, triggering cancer. Senescence (the slowing down of the cells' ability to divide and grow), malignancy (tumour growth) and degeneration (cell decline) mark the end of a life powered by the unwaveringly continual processes of DNA transcription, RNA translation and protein synthesis.

▶ Telomeres, the regions at the ends of chromosomes, shorten with age.

Human genetic variation

Every human being has the same set of proteins. We all have haemoglobin in our red blood cells and type-I collagen in our skin and bones. So why are we not all exactly the same as each other? For that we have to thank human genetic variation.

While it is essential for certain sections of the genetic code to be transcribed and translated error-free, for the blueprint to be followed to the letter, the genetic code contains a whole catalogue of variants, as well.

These variants are not damaging to proteins, and, with the exception of identical twins, they make us all different. They are distributed throughout the genetic code and may include modifications that involve anything from tiny changes at a single base to alterations involving whole sections of DNA. In fact, scientists have calculated that we each have roughly between 4 and 5 million variants. We know this because major advances in sequencing mean we can now study whole genomes, comparing one person's genetic code to another's.

Variation in the genetic code enriches us as a species. It makes us individuals; affects the efficiency with which proteins work; shapes our appearance; influences our response to illnesses, defending us against disease or making us more susceptible to it. This variation extends to how our bodies respond to medication – for example, whether a particular medicine will cause serious side effects for us, or if your body's metabolism will make a specific antidote useless to us, forming the aspirational basis of personalised medicine.

So human genetic variation can bring about individual differences through influence on protein function. In evolutionary terms, this may have equipped our ancient predecessors with the ability to adapt to changes in the environment, giving some groups a selective advantage over others. Exploring this side of the human genome is helping unravel the previously untold history of human development and evolution. On a day-to-day basis, though, it is transforming modern medicine and helping to solve crimes.

◀ **Human genetic variation is unique to an individual, except for identical twins.**

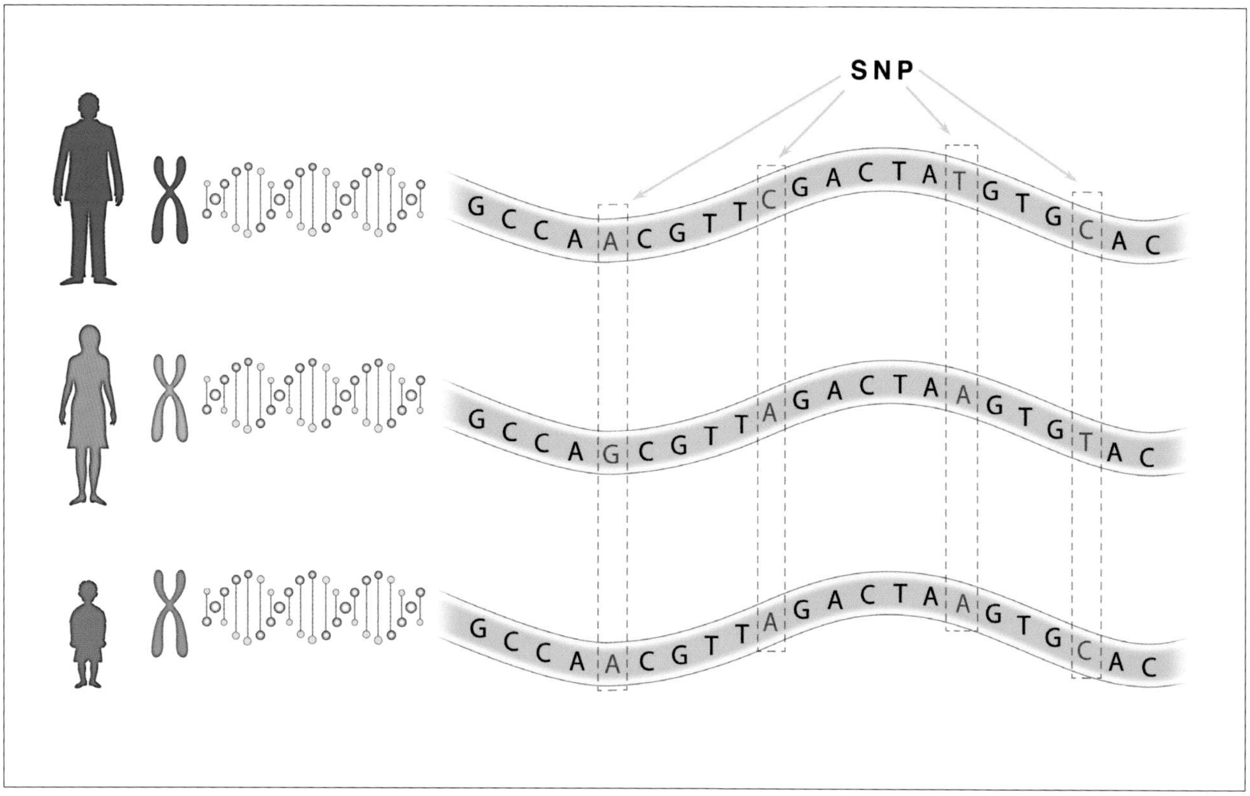

▲ **Single nucleotide polymorphisms (SNPs) occur when different bases are found at a particular locus in different people.**

SNPs ('snips')

The term single nucleotide polymorphism (SNP) refers to different forms (polymorphism) of the base at a particular position, or locus, in the genome. It might mean that where one person has 'A', someone else has 'T'. SNPs occur once every 1,000 or so bases. They are usually found in regions of the genetic code that are not transcribed, but may occur in coding regions too, so have the capacity to influence the efficiency of protein function.

SNPs probably do not have a direct effect on our health, but we can use them to study certain common, complex medical conditions. For example, the detection of a certain combination of SNPs, so-called disease markers, in some people has been associated with a higher chance of having the inflammatory bowel diseases, ulcerative colitis and Crohn's disease. The research studies that have mapped thousands of SNPs across the genome are known as genome-wide association studies (GWAS) and they have been used to help advance our understanding of many other conditions in which the cause is suspected, at least in part, to be genetic.

People who are related are more likely to share certain combinations of SNPs across their genetic code, as are individuals from a similar ancestral origin. We can exploit this observation in academic research to study the molecular past of our species, and in genetic testing to reveal ancestry and help people find unknown relatives.

▼ **Subtle changes in the genetic code shared by relatives can explain familial resemblance.**

Copy number changes and structural variation

With SNPs there is no alteration in the number of bases. However, copy number changes and structural variation result in the addition or removal of DNA sequences. There are thought to be around 2,500 variants of this kind across the genome, all of them contained within non-coding DNA. Again, these do not appear to affect health, but probably influence the expression of some genes, especially those under selective evolutionary pressure.

The default number of copies of any particular part of the genetic code is two – one from each parent (1+1 = 2). Copy number variation (CNV) refers to a change from that default. Changes are such things as missing runs of DNA sequence (deletions) or additional copies (insertion). Collectively these are known as 'indels'.

Structural variation on the other hand is thought to occur as a result of remnants of insertions of viral DNA that occurred over thousands of years ago. Rather than having an adverse effect on health, these remnants appear to have enhanced human adaptation, conferring selective advantage. Much of this variation was not thought to have any influence on our molecular workings today, which is why the moniker 'junk' is so dubious. In reality, though, our understanding of its true effect on the way we work is only just coming to light. It is likely to be much more influential than any notion of 'junk' suggests.

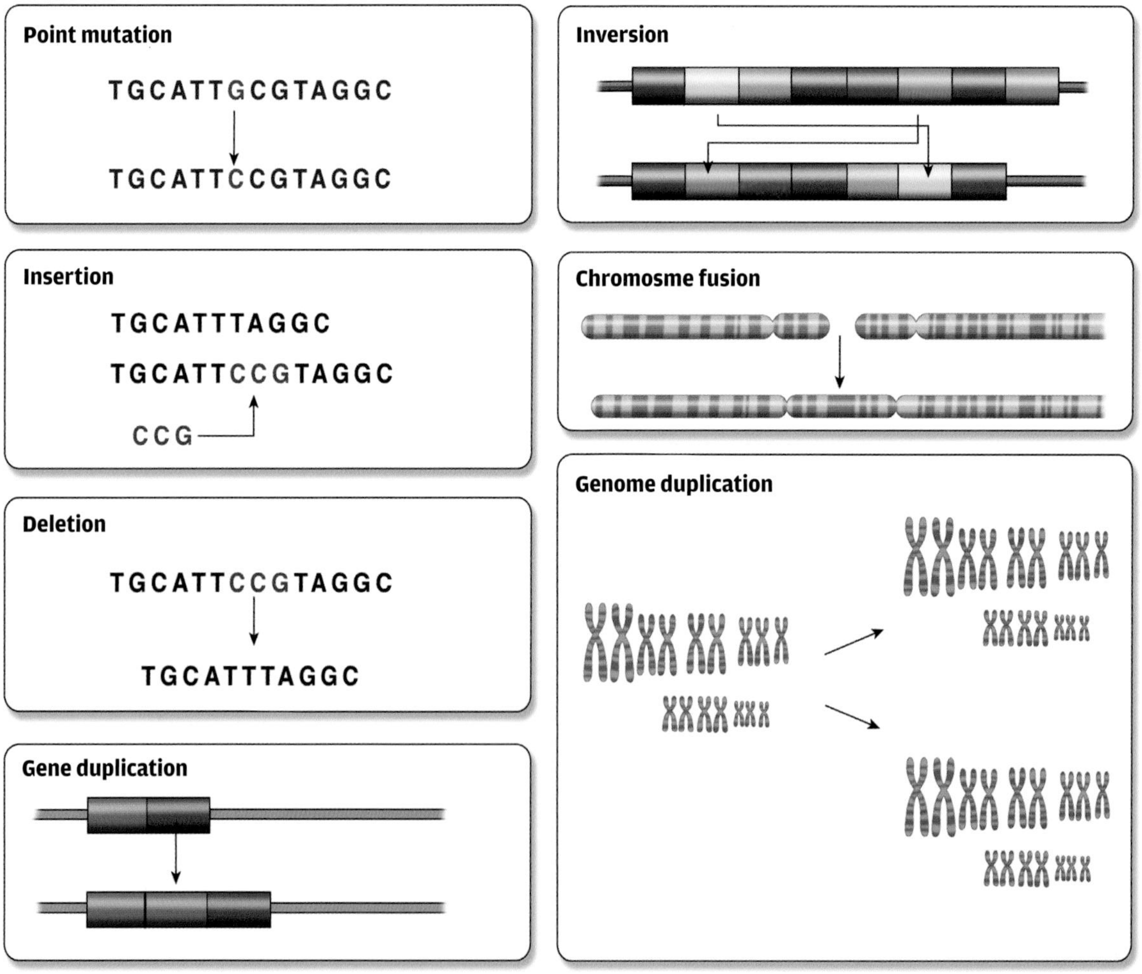

▲ **Point mutations, or changes at a single base, do not affect the number of bases like other changes in the genetic code can.**

How to predict pathogenicity

There are several different software programs that have been developed to predict pathogenicity – something's ability to cause disease. These take into consideration the potential effect the variant might have on protein function. Termed *in silico* analyses, these computerised tools incorporate many different factors that could be relevant in calculating whether or not a variant is disease-causing. Examples of the different software programs in routine use include SIFT (Sorting Intolerant From Tolerant), PolyPhen-2 and MutationTaster-2 and CADD (Combined Annotation Dependent Depletion). This is how class-3 VUS can either be reclassified as an innocent change or implicated in being medically important.

Pathogenic variants

There are some genetic variants that the body won't tolerate and so are much more likely to have a negative impact on our health. When the effect of the variant is so damaging that the constructed protein is faulty or can't be synthesised at all, it is described as pathogenic. It is this type of variation that causes genetic disease. Scientists have described more than 7,000 pathogenic variants so far. These variants affect 1 in 17 people, so individually seem rare, but collectively are, in fact, common.

Variants are more likely to be pathogenic when they partly or entirely disrupt protein synthesis. If the variant means one base is swapped for another – a C for a T, for example – the wrong amino acid could be introduced into the protein, which could completely mess up the way the protein works. This variant is known as a missense variant. Similarly a variant can mistakenly introduce a stop signal (nonsense variant) into the gene, altering which sections of the gene are transcribed and translated. 'Junk' DNA could be transcribed if the whole genetic reading frame is altered by an unfortunately placed variant (frameshift), drastically changing the protein we end up with. In the same way, deleting some or even all of a gene can also damage the constructed protein.

Transcription factors – the proteins that prompt gene expression – can also contain pathogenic variants. If this happens, not enough of the protein is produced.

In summary, then, damaging variants can exert their effects either by creating an abnormal protein (poor quality) or by not synthesising enough protein (wrong quantity). If you also take into consideration epigenetic factors that switch genes on or off, you have another molecular mechanism that can alter gene expression and therefore protein production.

▼ **Pathogenic variation affects the structure and function of key proteins on the body.**

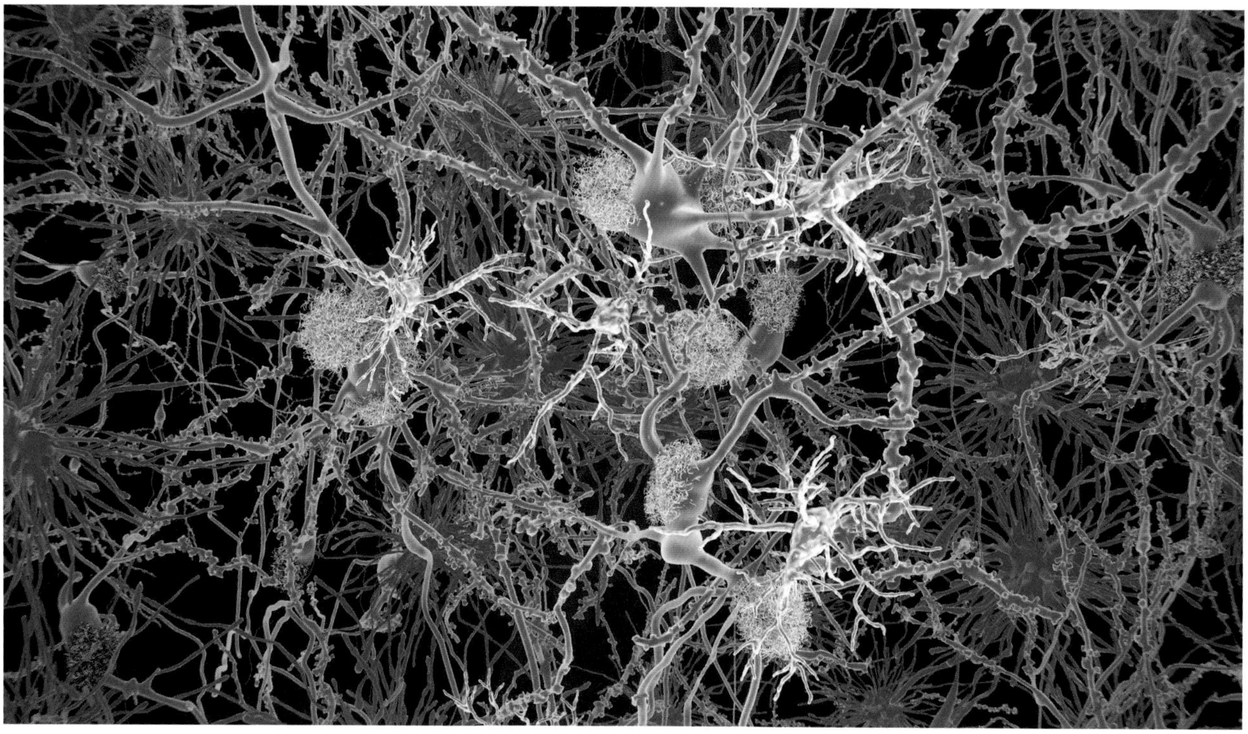

The spectrum of variation

From the inconsequential to the downright damaging, and from curiosities in the molecular archive to enhancing elements of gene expression, variation presents a full spectrum of effects. In the era when detailed genome sequencing is becoming commonplace, thousands of genetic variants are routinely detected.

To make some sense of this spectrum, scientists have developed a classification system that can help medics understand what's happening in order to provide focused treatment. A joint enterprise between the American College of Medical Genetics and Genomics and the Association for Molecular Pathology, the system categorises variants based on their pathogenic potential with the aim of identifying those with medical relevance, especially in the diagnosis of rare genetic diseases. In the American College guidelines, innocent changes in the code with no predicted damaging effects are deemed as class 1. At the other end of the pathogenicity spectrum is class 5, the rare variants that occur in the protein-coding regions, especially in sequences that are so intolerant of change that the code has remained exactly the same across many species. In the middle sits class 3, variants of unknown significance (VUS), a decision-making fence to sit on and testament to the fact that the more we discover about the human genome, the more we still have to decipher.

Mutations have mutated

Understanding the notion of genetic variation means that scientists no longer use the term 'mutation' when referring to disease-causing or deleterious changes in the genetic code. All changes could, by definition at least, be deemed 'mutations', but we now know that the vast majority of genetic variation is not, in fact, damaging. Thinking of these as variations, rather than mutations, helps us instantly acknowledge that changes are not always bad for us.

My brother, the banana

How similar is my DNA to the DNA in one of my first-degree relatives, such as my parents or my children? If it's only 50% the same, then how come the DNA sequence we share with a banana is 60%?

Our parents each give us 50% of their DNA sequence. The DNA that goes into the egg and sperm is a combination of theirs and has gone through a process of mixing and reforming, and then recombining at meiosis. This means that the full genetic sequence our parents pass on to us is different every time, so our DNA is different from our siblings', unless one of those siblings happens to be an identical twin.

When it is said that our DNA is 99.9% the same for every person, this refers to the number and type of genes that we have in common, as opposed to the similarity of the genetic code itself. As a species, humans have the same number of genes and make the same proteins as each other. This is different to chimpanzees (98.5% similar), bananas (60%) and fruit flies (47%).

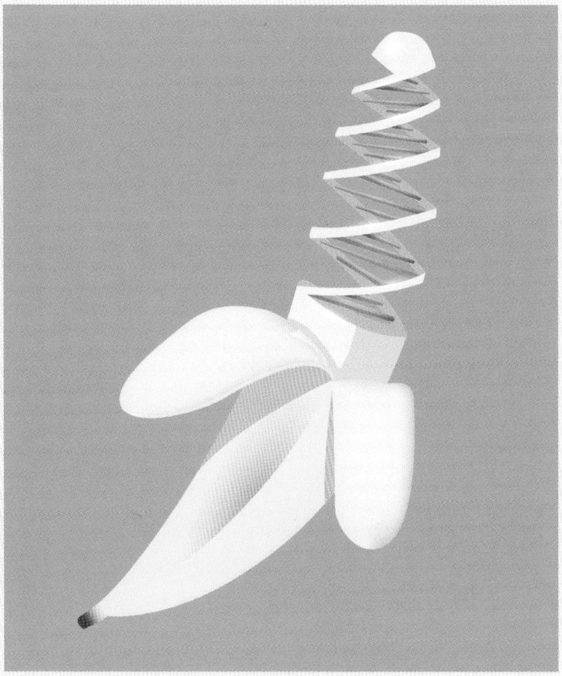

Human physical variation

Studies of the whole genome, including the 'junk' DNA, has given us valuable insight into what makes us all different. None of us looks the same – even identical twins display differences (although perhaps only to their mother). Childhood growth charts and school class photos illustrate that there is a whole range of heights deemed healthy and acceptable for any given age. Why? Because of those subtle variants in the genetic code.

Studies of different markers, variants across the genome, have identified a handful of SNPs that influence height. Mostly, these markers have been identified in people who are considerably below average height, but still within a statistically normal distribution. The variants are not pathogenic, they do not damage any proteins, but they seem to have some influence upon growth. They are often found in close proximity to growth genes, which, when a pathogenic variant disrupts their function, can result in dwarfism, for example.

Scientists have discovered genetic factors that determine the shape of noses and chins. We have already encountered one of these genes, *RUNX2*, in its capacity as a trigger for type-I collagen gene expression in bones. Variants functioning as genetic markers in *RUNX2* have also been linked with how wide and pointed our noses are. Subtle variants in another gene, *EDAR*, have been linked to how much our chins protrude. There will likely be many other markers that we can identify as lying behind physical variation, each with different potential for scientific impact.

▼ **Genetic variation can create physical variation, such as in height.**

Making sense of it all

The human genome is complex. Fortunately, we have 21st-century technology at our fingertips, so we can start to make sense of it all. Using computer science and internet-based technology, as well as an ever-improving understanding of the field, we can sequence, analyse, interpret and store the data we extract from the genetic code. With international consensus ensuring there is standardised practice across the globe, we now have what it takes to unlock all the secrets held within the code.

Naming genes

The naming process is conducted in accordance with regulations set out by a specially appointed committee: The Human Genome Organisation Gene Nomenclature Committee (HGNC). They name genes after the function of the protein they encode. For example, *COL1A1* refers to collagen I, the condition with which pathogenic variants are associated, such as neurofibromin in the neurocutaneous condition affecting the nerves and skin, neurofibromatosis, or eponymously named syndromes linked with pathogenic variants in the gene, for example, Emerin after Emery-

▶ **Making sense of the human genome has been facilitated by radically advanced technology.**

Lunatic fringe

Before the creation of the Human Genome Organisation Gene Nomenclature Committee (HGNC) some fun was had with gene naming. One way to study a developmental role of proteins is to deliberately disrupt the genes that make them and see the effects. In the case of the fruit fly, *Drosophila melanogaster*, this led to some rather whacky gene names.

Scientists conducting experiments to remove or 'knock out' certain proteins during early *Drosophila* development noticed unusual characteristics in the resulting flies and named them after their appearance: Lunatic Fringe and Indian Hedgehog. When another researcher in the same field was looking for a name to call the gene he had discovered, he decided on 'Sonic Hedgehog' in homage to the computer game his kids were into at the time.

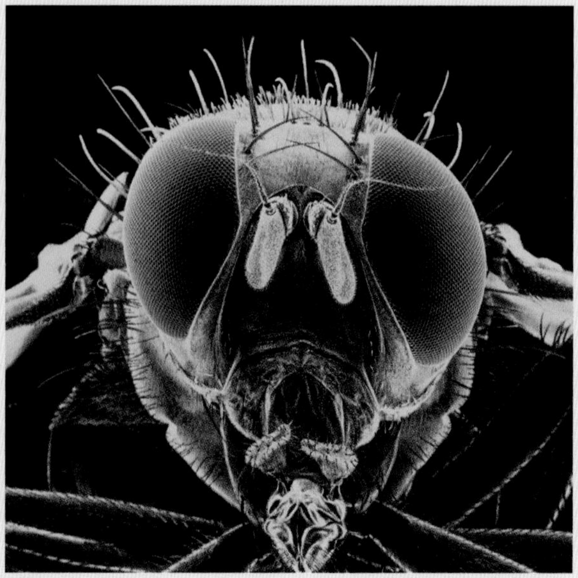

Dreifuss muscular dystrophy. They are also represented by an approved abbreviation. Using a standardised system in this way removes the potential for confusion.

Genome builds and browsers

Genome browsers are web-based tools that provide access to information about the genomes of different species. They act as a reference library for the entire genetic code, accessible to researchers and scientists who can use these browsers to help them analyse and interpret genetic data. We can use them in many different ways, including to identify highly variable regions, as well as rare, pathogenic variants with medical relevance, and to compare the same gene across different species to find functional similarities and explore evolutionary conservation.

It was the Human Genome Project that led to the creation of a reference genome, the gold standard against which other DNA sequences can be compared. In scientific terms, using the same reference genome ensures consistency from one laboratory to the next – in other words, comparing apples with apples and not pears.

The reference genome is never fixed, though, and needs regular updating and curating with new discoveries and better understanding of the genetic code. The latest version, the *Genome Reference Consortium human genome build 38 (GRCh38)* was published in December 2017. Anyone can access the reference genome using the human genome browsers Ensembl (www.ensembl.org) and University of California Santa Cruz; https://genome.ucsc.edu. These browsers have many other uses, and play an essential role.

▼ Technology helps build up a picture of the human genome that can be used by scientists to explore the mysteries of genetic code.

Bioinformatic workflow

Analyse:
First, genomic data is analysed to identify all the different variants within an individual's genetic code. This is known as variant calling. Scientists need genome databases to provide reference that enables them to find out if a variation is known to occur at any one particular spot in the DNA or genetic locus, and also to make sure there is consistency in the names given to each variant. Annotating these variants in the same way across different laboratories is essential for comparing genomes in a meaningful and useful way.

↓

Filter:
The next step is to filter out and remove the variants that are not of interest. Using the example of rare-disease variants, the bioinformatic filter is set to exclude ones that are so common in the population they cannot possibly be damaging. Ideally, this step results in a much shorter list of variants for the interpretation step.

↓

Interpret:
More database resources are required here, bringing together further information about the genes in which the remaining variants have been found, the function of the proteins they code for, and the potential damaging effects that the variants introduce. Bioinformatic pipelines are constantly being redesigned and redeveloped to improve the interpretation of genomic data.

↓

Now we just have to store all this information

Bioinformatics explained

The advent of high throughput sequencing brought with it the fresh challenge of working out how to handle all the data that reading the genetic code generates. Bioinformatics was originally developed to apply computer science to biological information. It can also be applied to genomic data. This requires bringing together different resources, such as reference databases and software tools, to manage the 3 billion bases and thousands of variants identified every time a genetic code is sequenced. By combining computational analysis with biological knowledge, bioinformaticians can delve deep into the genome to seek out the regions that interest them. For example, in rare-disease genetics, there could be just one pathogenic variant in the whole of the genome that is causing a medical condition. Getting from 3 billion bases down to that particular one calls for a bioinformatic pipeline or workflow process to apply the right set of rules for the task.

Big data

Data are units of information collected in an electronic form for a computer to store and use. 'Big data' is information that is both high in volume and extremely complex, so conventional storage and methods of analysis are completely inadequate. Take, for example, a file containing the read-out from one person's genetic code. These 3 billion bases need 100 gigabytes of computer storage space – that is, 10,000 billion units of storage. Replicate this many times over, and this presents a significant logistical challenge.

The world of big data is underpinned and defined by the five Vs: volume (the amount of data), velocity (the rate at which it is received), variety (the many different types of data), veracity (its trustworthiness) and value (the potential for uncovering useful information). Technological advances improve computer analysis programs and reduce the costs of handling big data on a large scale. Linking bioinformatic information with other sources of big data, such as electronic health records, in pursuit of offering a truly personalised medicine service, presents the next logistical challenge in storage.

Gene patenting

To some, after discovering genes and their function, it is open season for exerting rights over their future use. Gene patenting is the legal granting of exclusive rights to a specific gene of the individual or institution who first identified it. Those rights relate to how that gene is used – for example, in diagnostic testing, scientific research or commercial exploitation – and can be in place for 20 years.

In 2013, in the USA, after more than 4,000 gene patents had been granted, the American Supreme Court ruled that, in fact, genes could not be patented. In Europe, however, gene patenting is still possible, as long as the gene identified is shown to have a function that was not previously known and that it has a potential use – a medical application, perhaps, such as the development of life-saving drugs.

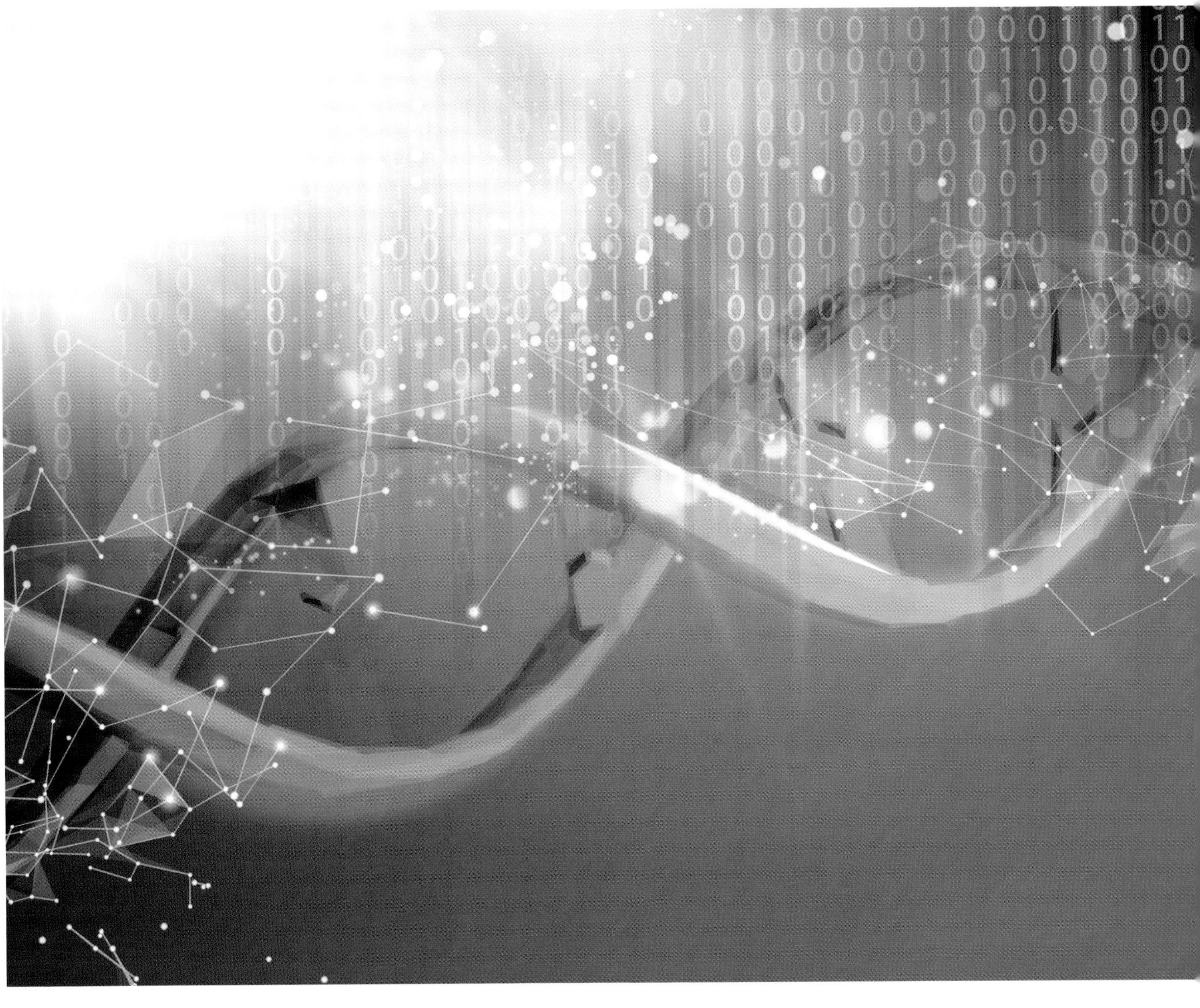

The science of -omics

Genomics encompasses every aspect of the genetic code. It relates not just to the parts that encode proteins, but absolutely everything about it, from start to finish, no holds barred: its structure, its function, its variation, its 'junk', its evolution, its map, its organsiation – the complete catalogue. From this, a number of other biological libraries have developed, offshoots of the genomic arena, together known as the *–omics*.

Transcriptome

The catalogue of all mRNA transcripts, the first read-outs from the genetic code in the protein-making process, is known as the transcriptome. It is made up of all the magazine articles without the interspersed ads, these having been spliced out, leaving just the essential reading matter. The transcriptome refers to all the mRNA sequences that one particular cell produces or to a group of specialist cells from a certain tissue,

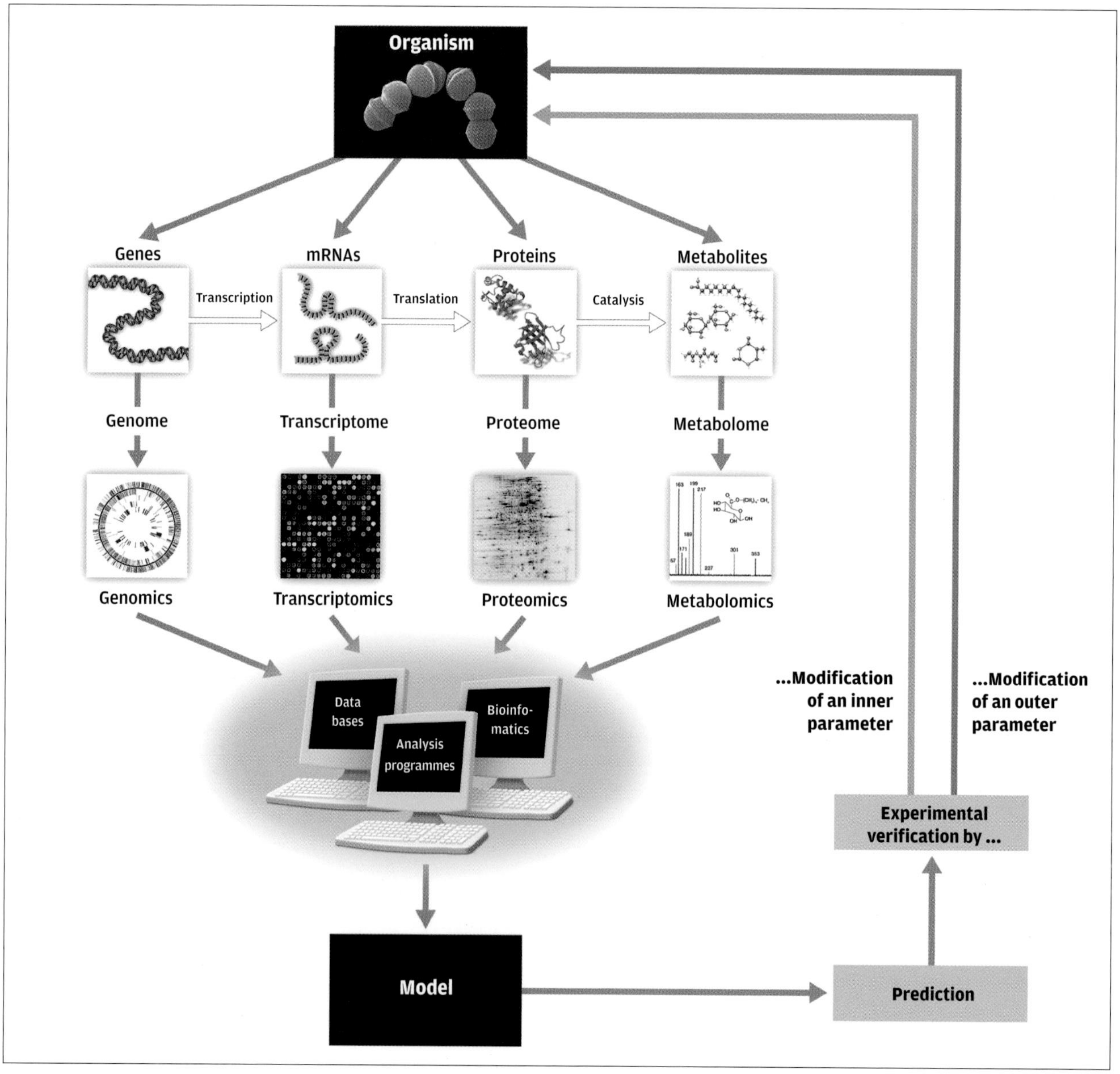

or it relates to all of the transcripts an organism produces – a human being, for example. It constitutes the essential first read-out of the genetic code, the sequences that then go on to make the proteins.

Proteome

The proteome is the complete collection of the proteins the body synthesises – several thousand of them. There are 20,000 genes in the human genome, but there are three or four times as many proteins in the human proteome. One explanation for this is that a single gene can produce a slightly different form of the same protein depending where it is expressed and what it is required to do. The final product might be shorter and more compact in one tissue compared to another, something that is achieved through a process called alternative splicing. This allows similar proteins to be slightly different, usually in length, adapting to the needs of the tissue's environment.

Metabolome

A sub-group of the proteomic collection is the metabolome. This refers specifically to the group of proteins concerned with running the body's metabolic system, all of the chemical compounds that make up the pathways and enzymatic processes that make us function at a cellular level. Studying the metabolome helps us to understand how our cells work and what happens when a single element goes wrong, the consequence of a genetic disorder causing an inborn error of metabolism, for example.

Variome

Variome studies focus on the thousands of differences from one person to the next that make up normal human genetic variation. Improvements in genomic sequencing technologies have opened our eyes to the vast array of differences we each have throughout our genetic codes. Why we have all these differences, what they mean and where they originated are increasingly of scientific interest. International collaborative efforts, such as Human Variome Project and the International Hapmap Project, seek to answer these questions. The Database of Genomic Variants is collating all of the identified variants to build up a full picture of the scale of human genetic variation – a major undertaking.

Epigenome

The epigenome is a collection of all the epigenetic markers and chemical modifications to the genetic code, the ones that are not heritable. The fascinating concept of this is that it changes over time, so that one individual's epigenome will not remain the same throughout their lives, but will vary, responding to exposure to environmental factors and the ageing process. This branch of biological science will, as it continues to evolve, give us valuable insights into the effects on the genetic code of us simply living our lives and just how exactly nature interplays with nurture.

Ciliome

We are really getting down to the small print now. The proteins found in a particular cell organelle called the cilium are collectively known as the ciliome. Cilia are tail-like extensions protruding from the cell membrane of some specialist cells. Such ciliated cells are found in the retina at the back of the eye, some types of kidney cells, and cartilage-making cells in the joints and growth plates at the end of bones. They are also responsible for establishing the left- and right-side patterning of the early embryo.

When ciliated cells go wrong, foetuses can develop with the body organs on the wrong side of the body, situs inversus, literally meaning that everything is in the opposite place. A group of rare genetic disorders characterised by abnormal cilia function, called ciliopathy, has emerged over recent years. As a result of researchers building up such in-depth knowledge of conditions like these, we can now develop targeted treatments to help those affected to overcome them.

▼ The stalk-like projections of this cell, seen under an electron microscope, are the cilia that sense what is happening outside the cell.

АКАДЕМИЯ
НАУК

CHAPTER 3

Working with ancient DNA

Unravelling the history of the human species has always been the domain of archaeologists and anthropologists who meticulously collect and curate artifacts and study the remains of our ancient ancestors. An historical record emerges, a combination of all the available evidence, interpreted so that we can imagine a picture of the past.

DNA and archaeology

While interpreting archaeological findings can give us invaluable insight into the past – the tools our ancient ancestors used, for example, even fragments of their bones and teeth – filling in the gaps requires a level of creativity that belies the archaeological science that goes into deciphering these ancient clues. But, when DNA and archaeology come together, the earliest beginnings of humankind begin to reveal themselves, often in untold and unexpected ways.

DNA and the fossil record

Artifacts – such as arrowheads and flints, pottery and earthenware materials, jewellery and coins – reveal ancient practices and show us what life might have been like for our ancestors. Finding bones and teeth that are thousands of years old enables archaeologists to study the physical features of their owners – and from them we can deduce such things as height, skull formation and face shape, which we can then compare to the equivalent features in modern-day humans.

Observing the size and shape of teeth can give us an idea of what our ancestors' diet was like – was it plant-based or carnivorous, for example? Disruption to the bones or defects in them might reveal cause of death from injury or disease, providing insights into the dangers of life gone by. Combining all of this information builds us a picture of how humankind established itself, and developed and evolved over time.

Radiocarbon dating

Archaeological study involves a lot of educated guesswork, though. What if there were a foolproof scientific method that we could apply to the historical record, supporting archaeological conclusions or filling in the gaps?

There is. Radiocarbon dating enables us to tell the age of ancient objects that contain organic matter that has been derived from plant or animal material. Components of living things contain different forms of carbon called isotopes. One of those isotopes is called carbon-14, which all living things constantly acquire from the environment during life. After death, however, carbon-14 slowly disintegrates through a process called radioactive decay.

Radioactive decay is predictable. Working out how much carbon-14 is still contained within something therefore gives us a way to estimate how old that thing is – as long as, that is,

▼ The slow and painstaking archaeological excavation of a human skull.

▼ Combining newer scientific methods with archaeological findings is a powerful mix.

the thing is less than 50,000 years old. This is because once an object is more than 50,000 years old, radiocarbon dating is no longer accurate.

Combining the arts

So, science can support archaeology to work out the age of remains that contain organic matter, such as skeletons and teeth from, for example, Neanderthal man. But what if we could use those biological specimens in other ways to give up the secrets of the past? Perhaps they would bring us closer to the genetic codes of our ancient ancestors. Combining this information with archaeological findings, we could not only compare the genetic make-up of ancient humankind to our own to delve deep into its very being, but also start to work out where we came from, and, perhaps, where we may be heading towards.

Discovering Neanderthal man

The Neanderthals are extinct early humans that lived in Europe and Asia 400,000–40,000 years ago. They are so called because in 1856, the first specimens – 16 bone fragments and a skull – were discovered by quarrymen in the Neandertal, a small valley along the Düssel River in Germany's Rhine region. These remains were quickly recognised as belonging to a human population different to our own. After the discovery, experts realised that fossils found earlier in the 19th century in Belgium and Gibraltar were Neanderthal too.

Since then more than 400 skeletons have been discovered, located as far west as Figueira Brava on the Portuguese coast, spreading easterly to Iraq, south to Somalia and north to Siberia. There are concentrated pockets of discovery in France and in Krapina, Croatia. Mostly these remains, along with other Neanderthal artifacts, have been found in caves – giving us our colloquial term, 'cavemen'.

The missing link?

On an evolutionary trail in which we share a common ancestor some 500,000 years ago, these ancient humans are considered to be our closest ancestors. In 1859, three years after Neanderthals were first discovered, Charles Darwin published *On the Origin of Species*, revealing his theory that populations evolve through a process of natural selection. Imagining a prehistoric world in which these ancient relatives of ours roamed requires only a creative mind, as does linking their existence with our emergence as modern-day man. But what if we had a way to travel back through time and bring the past to the present using modern technology?

▶ A striking reconstruction of a Neanderthal man, based on the most recent archaeological information.

DNA digging

The ability to unlock the secrets of time has emerged in the shape of the genetic code: after all, DNA lies in the bones and teeth of the Neanderthal just as it does in our own. If we could resolve the thorny problem of getting it out of there, we could apply the technological advances of the day to sequence it. That way we could compare the read-out with our own genomes, looking for similarities and for differences. It would create a kind of map, charting the path from our ancient ancestors to modern-day human.

Extracting ancient DNA

Extracting ancient DNA was not without its major methodological headaches, which scientists had to work hard to overcome. While DNA is a very stable molecule with a long shelf life, it does degrade over time. The long strands of DNA that contain the 3 billion base pairs – A, C, T and G – fragment and break up, leaving big gaps in the code. When they were first discovered, some specimens of Neanderthal bones were stringently cleaned, reducing the amount of DNA available to modern scientists to study. Contamination of the specimens further complicated the problem: early tests revealed that much of the seemingly Neanderthal DNA was, in fact, microbial in origin, from the bacteria and other micro-organisms that inhabited the ground where the bones were excavated. It even contained DNA belonging to the laboratory scientists themselves. When we are talking about such tiny concentrations of DNA, even miniscule contaminants from the skin or hair of modern humankind can overwhelm the Neanderthal's.

So, specialists in ancient remains have developed the skills and methods to extract DNA using specially designed facilities to prevent such contamination. They implement quality control checks to make sure the DNA is pure. Furthermore, they have identified which sections of bone are most likely to give up DNA. Now, scientists have managed to extract thousands of specimens from all sorts of ancient remains for genetic testing.

▼ Archaeologists examine bone fossils at a Neanderthal excavation site.

How to extract DNA in your kitchen

You can extract DNA yourself, for example from fruit, using objects commonly found around the home. DNA is contained within the seeds of fruit. This experiment uses strawberries, but kiwi fruit and banana will work just as well. All you need to do is to release the DNA from inside the cells and it will appear before your very eyes – the stuff of life!

You will need

3–4 ripe strawberries
1 ziplock bag
2 glass containers
2 teaspoons washing-up liquid
½ teaspoon salt
120ml water
1 coffee filter
Rubbing alcohol (also called isopropanol)
1 cocktail stick or bamboo skewer

Method

1 Place the strawberries into the ziplock bag, gently squeezing out the air as you do so (just to prevent the bag from popping open), and seal it closed.

2 Smash up the strawberries with your hands. This breaks open the cells in the fruit, making it easier to release the DNA.

3 Take one of the glass containers and in it gently stir together the washing-up liquid, salt and water. This makes a lysis solution that will break open the cells and free the DNA inside.

4 Add this solution to the bag of squished strawberries and seal it again. Gently mix the solution around with the fruit, being careful not to make a lot of bubbles. Now the DNA is dissolved inside the messy liquid inside the bag.

Scientists apply the same methodological principle in scientific and medical genetics laboratories - using samples of blood and saliva in place of strawberries.

5 Place the coffee filter inside the second container. Pour the liquid from the bag through the filter. Pull the edges of the filter up into a balloon shape and squeeze gently so that the liquid passes through, leaving the strawberry mess behind in the filter.

6 Now to get the DNA inside the container out of the solution. Double the volume of liquid inside the cup by adding the same amount of rubbing alcohol as you have solution. You should see two layers form inside the cup. The one at the bottom has a strawberry-coloured tinge to it; the upper layer is white and slimy.

7 Using the cocktail stick or a bamboo skewer, pick up the slime from the top layer. Behold – this is purified DNA!

Our genetic history

The humble beginnings of the analysis and interpretation of ancient DNA were not straightforward. In recent years, though, important advances have developed in parallel with major improvements in the field of genomics as a whole. In its own evolutionary pathway, the focus has shifted from studying the DNA within the mitochondria and Y-chromosome to looking at much larger sequences contained within the genomic DNA.

Start with the blood

Originally scientists used the poor-quality, contaminated DNA extracted from ancient remains to study blood groups – classified as A, B, AB and O and known as ABO blood grouping. When scientists realised that subtle changes in the genes encoding haemoglobin could result in different proteins, ABO blood grouping at the genetic level formed the basis of the first studies of genetic variation. But this system is extremely limited, as there are so few blood types.

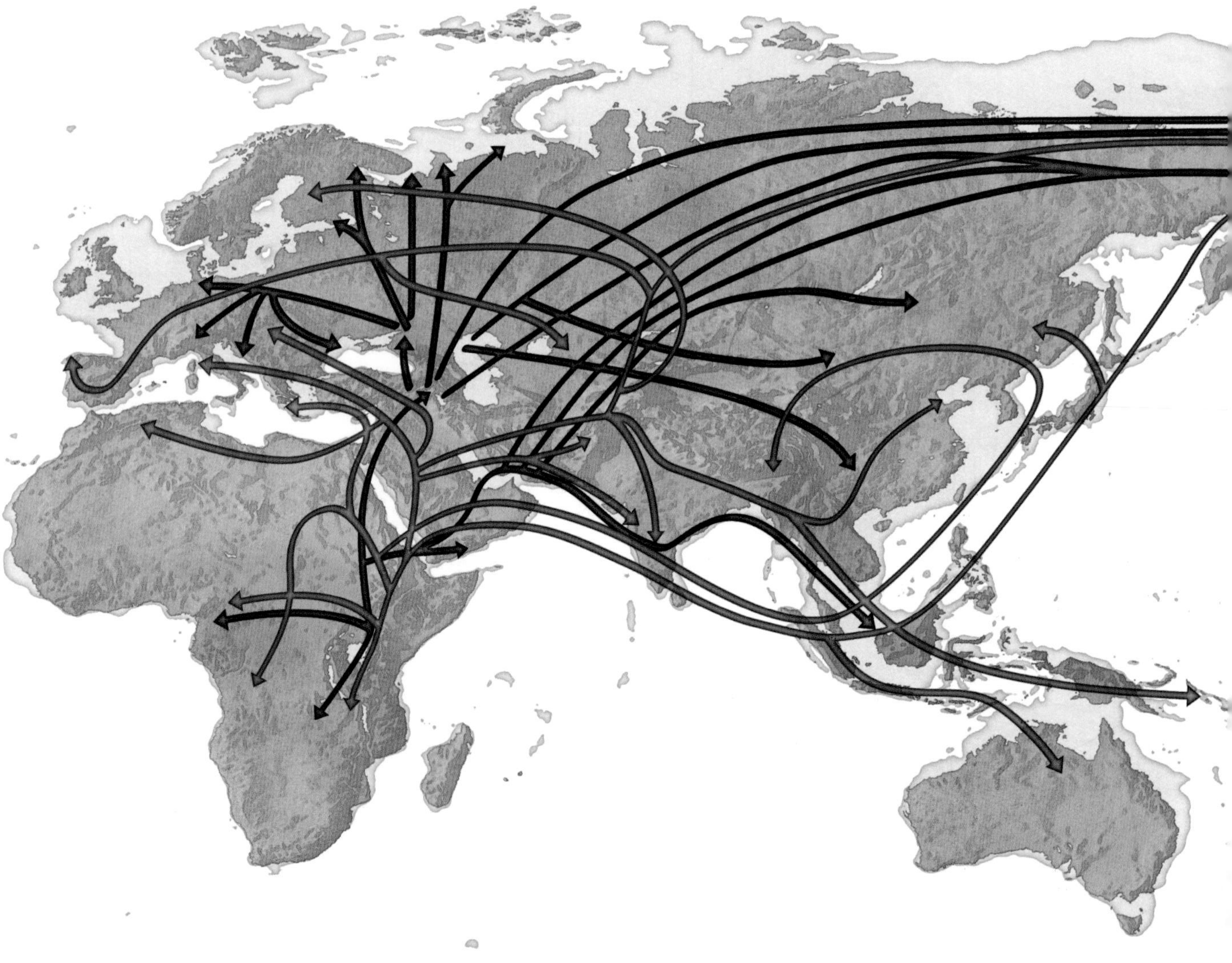

▼ The human genetic map – ancient migration according to the DNA record.

Studying mitochondrial and Y-chromosome DNA raised expectations, however. mitochondrial DNA is much less subject to random changes and so is transmitted down the female line generally unchanged. We are able to calculate the rate of accumulation of new changes. So, comparing mitochondrial DNA across populations can give an idea of a timeline, linking the origins of one population to another. This bore the notion of Mitochondrial Eve, a conceptualised being who would have existed between 166,000 and 200,000 years ago. Scientists have used the study of Y-chromosome variation in a similar way, greatly assisted by population-level testing and analysis of the mitochondrial DNA of 1 million people, through the National Geographical Society's 2005 Genographic Project.

What DNA can tell us about the past

With the breakthrough that came with major improvements in extracting higher-quality DNA from ancient remains also came challenges inherent to the DNA's age. DNA degrades, decays and fragments over time, leaving gaps in the code. But we can still use it to map genetic variation across the genome, letting us identify distinctive genetic signatures that people from a similar origin share. The fascinating developments in extracting ancient DNA more efficiently, mapping human genetic variation and then applying sophisticated mathematical formulae to release its full potential has opened up untold possibilities, transforming our understanding of human evolution and population migration.

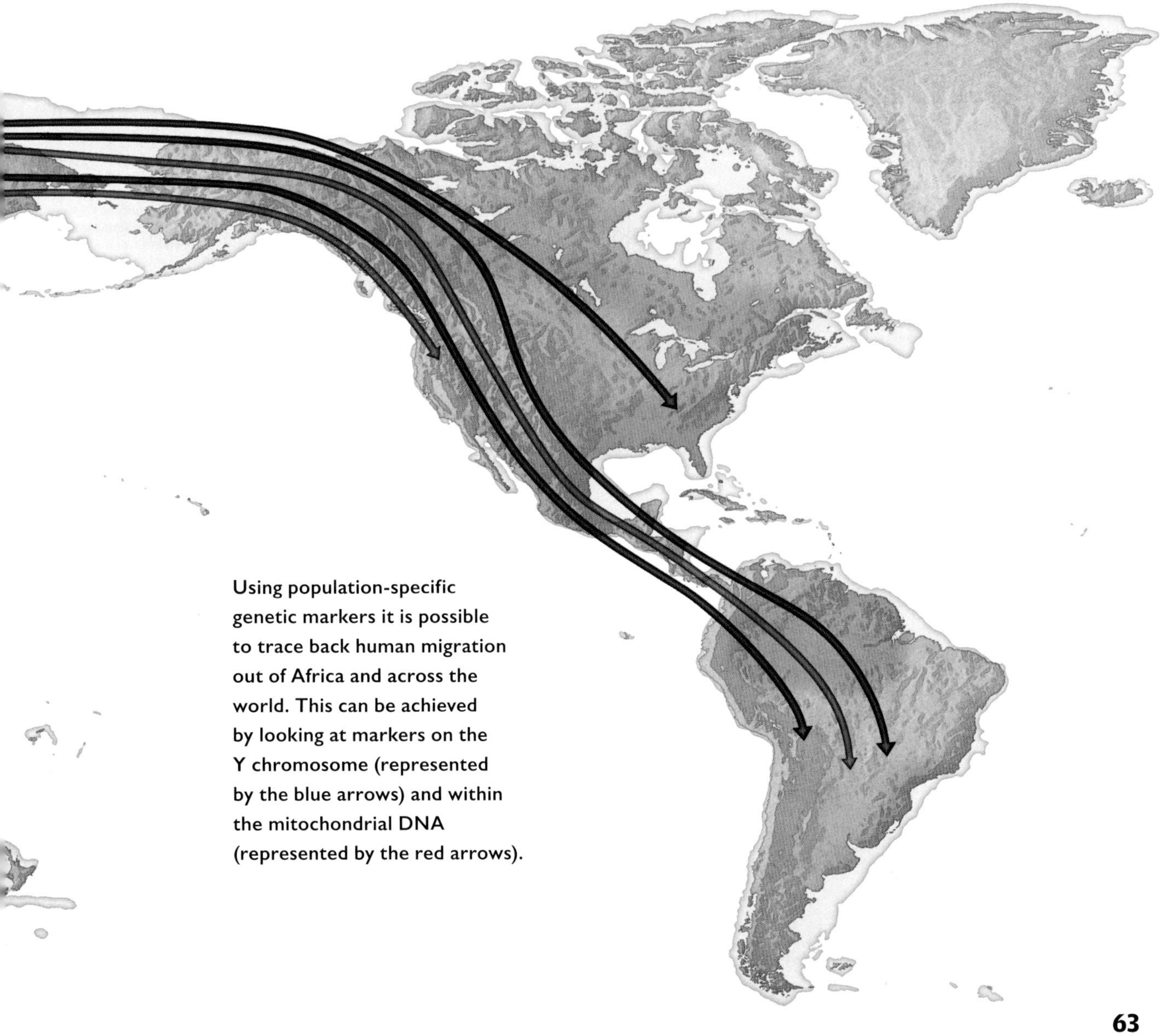

Using population-specific genetic markers it is possible to trace back human migration out of Africa and across the world. This can be achieved by looking at markers on the Y chromosome (represented by the blue arrows) and within the mitochondrial DNA (represented by the red arrows).

Using mitochochondrial DNA

In 1987, Rebecca Cann analysed the mitochondrial DNA of 147 people from different places around the world, such as Japan, South and East Africa, and Native America, to find out how we all related back to our most recent common ancestor. She tracked certain markers in the samples and matched them to where the people came from. She then compared them all, not only with each other, but also with markers in the mitochondrial DNA of the chimpanzee. Until her work was published, it was thought that *Homo sapiens* evolved from a common ancestor 2 million years ago, independently occurring in many different places across the Old World. This is the so-called multiregional theory. Cann's analysis of mitochondrial DNA revealed instead that all humans arose in Africa and shared their most common ancestor 200,000 ago.

Using Y-chromosome DNA

Y-chromosome DNA also presented two major deviations from previous theories: first, like Cann's findings, it identified that the multiregional notion was not accurate. And, second, also like Cann, it identified that it was not 2 million years ago that we shared the common ancestor, but much more recently than that. The studies using Y-chromosome DNA methods confirmed both Cann's findings – first, that humans arose in Africa, and, second, have a common ancestor dating from a mere 200,000 years ago.

A quick spot of revision

Before we go on, it is worth recapping some of the material covered in Part One, as it forms the basis of how we have used DNA to map our past.

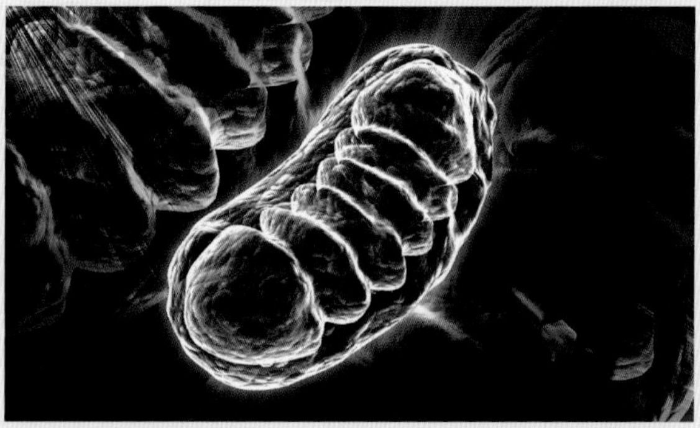

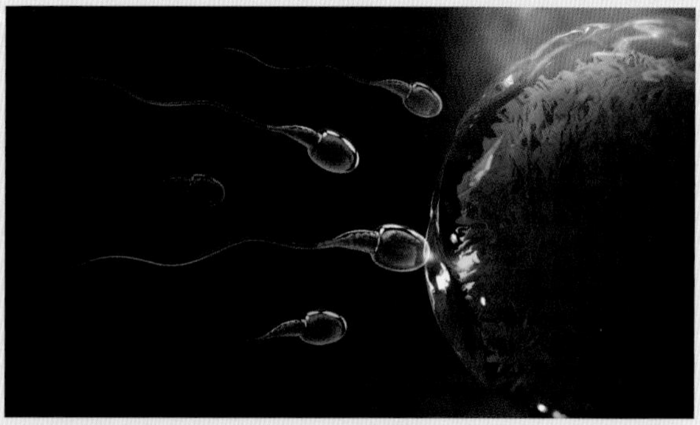

Mitchondria, those hundreds of powerhouses found in the cytoplasm of cells, contain their own segments of DNA. Much shorter than the DNA in the nucleus of a cell, it is almost exclusively inherited from the mother in her egg. Nuclear DNA is stored in the chromosomes and this is what makes up the genome. Just males have a Y-chromosome, so the DNA contained inside is passed only from father to son. Both males and females have chromosomes 1 to 22 and X, and this makes up the majority of genomic DNA.

To find similarities and differences between genetic codes, we look at the degree of genetic variation. These include the SNPs, single nucleotide polymorphisms, where one person might have a T and another a C base. These thousands of subtle changes usually do not affect the way a protein works, but are the result of accumulating random genetic changes in a population.

Each time an egg and sperm form, the cells that are responsible for gametogenesis undergo meiosis, the reduction division process that means 23 chromosomes are passed on from the mother and 23 from the father. Before the separation of chromosome pairs occurs, they come together to exchange material in a process called meiotic recombination. This means the chromosomes passed on to the next generation are not identical to one parent, but an amalgamation of the two.

To achieve this, breaks are made in the chromosomes, the exchange happens and then the chromosomes are resealed. In a phenomenon known as linkage disequilibrium, chunks of chromosome that are closer to each other are more likely to remain together on the newly created chromosome, whereas those further apart have a higher chance of not being passed down together.

The four-population test

The four-population test was developed by David Reich as a method of working out how closely related different populations based on shared genetic variants.

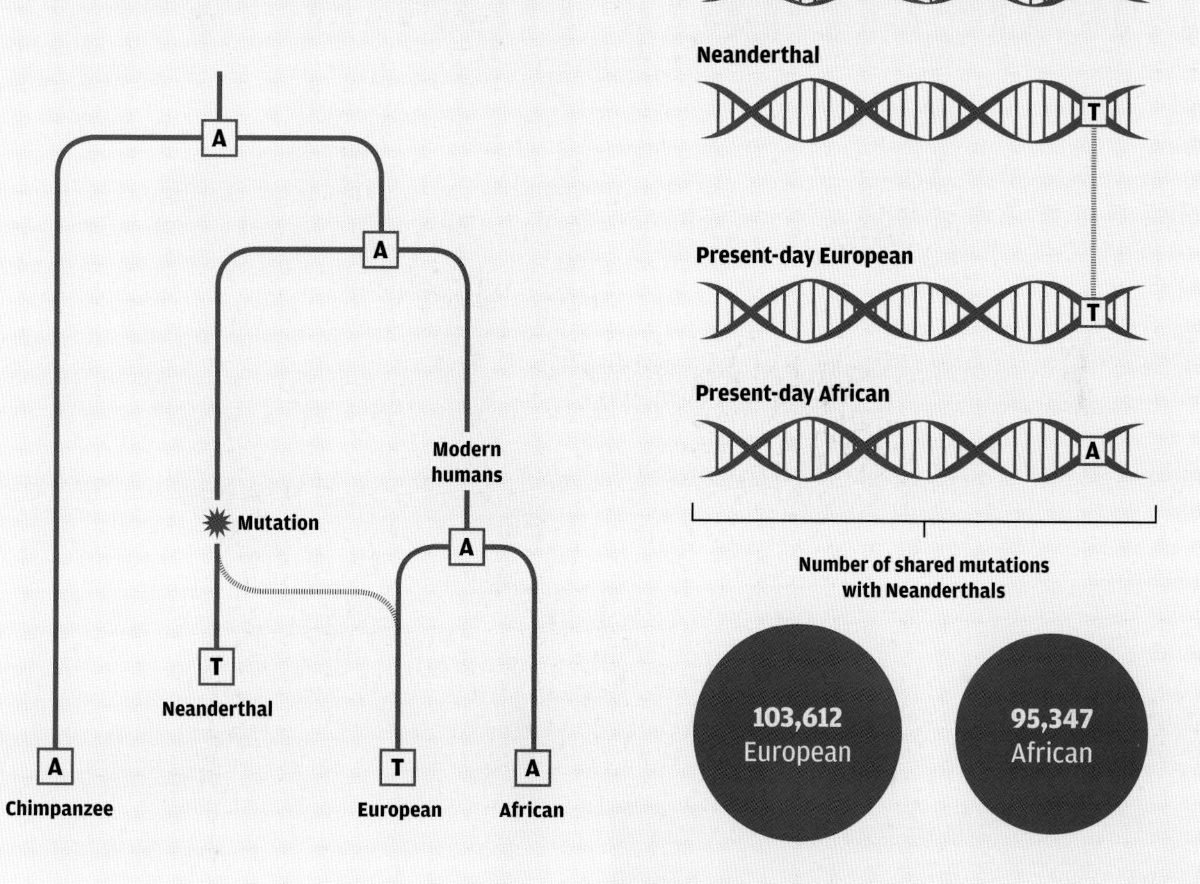

Genome mapping and the four-population test

Building up a map of the similarities and differences in the genome forms the basis of a branch of science called population genetics. People from a similar genetic background are more likely to share similar DNA sequences with each other compared with those from completely different backgrounds.

Mapping distinctive patterns of variation in different populations requires finding key SNPs at certain regions throughout the genome and identifying runs of DNA sequences that have been inherited together as a chunk without undergoing recombination – those regions of higher linkage disequilibrium. This means that we can create a map of different genomic signatures – markings characteristic of populations with a similar genetic background – and then study them in detail to find out more about their origins and their effects.

David Reich, pioneering geneticist in the study of ancient DNA from Harvard University, developed the four-population test as a basis for working out if different populations are related, by comparing their genetic maps. It was he, working in collaboration with Swedish geneticist Svante Pääbo and other leaders in the field, who revolutionised the extraction of DNA from ancient remains and who has devised mathematical models to interpret the complex clues held inside the genome.

In the four-population test, a series of genetic markers are compared between individuals from four different populations. Where, at a particular genomic location or locus, population 1 has an A; population 2, a T; population 3, a T; and population 4, a G, it is thought, or inferred, that populations 2 and 3 are more likely to share a common genetic background than populations 1 and 4. When this process is replicated over vast swathes of the genome, we can build up a signature map to represent each individual population.

The origins of modern-day humans

David Reich, his team and their international collaborators' groundbreaking work has transformed the field of human genetics. Studying the DNA of Neanderthal man enables us to begin to unravel all the questions archaeologists and anthropologists have about our links with the Neanderthals and our own ancestral origin.

Archaic humans

Modern human is from the species *Homo sapiens*. Fossil records reveal that there were other species before us – archaic humans. For example, fossils relating to *Homo erectus* have been found in multiple locations, including South Africa, China, Spain, Indonesia and Kenya. We believe that *Homo erectus* lived 1.8 million years ago, a significantly long time before *Homo sapiens*, a mere 200,000 years ago. Studies of the skeletons of *Homo erectus* show that this archaic species walked on two legs, like us, and had a similar body size, but that its head was larger than an even earlier ancestor – the 3.2 million-year-old skeleton called Lucy found in Ethiopa, Africa, in 1974.

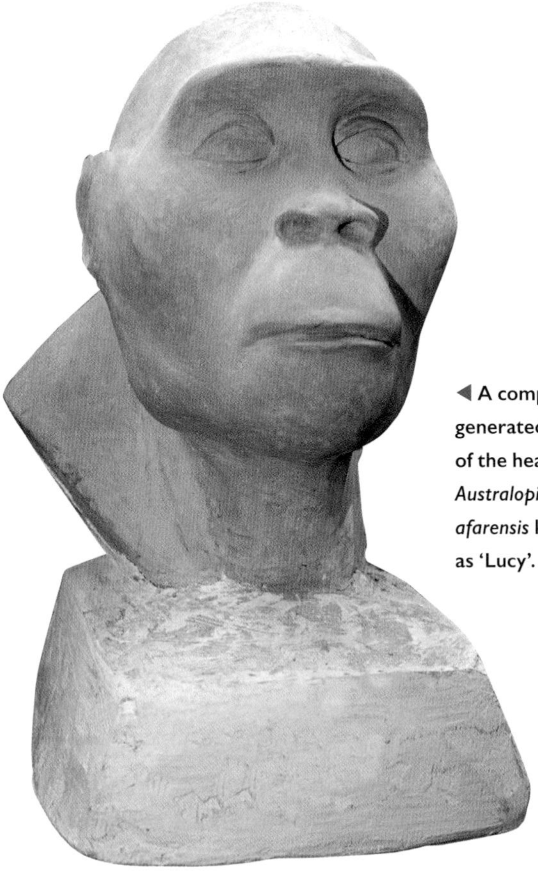

◀ A computer-generated model of the head of the *Australopithecus afarensis* known as 'Lucy'.

▼ The fossilised skeleton of a *Homo erectus* boy found in Kenya, Africa.

Lucy belongs to the species *Australopithecus afarensis*, which was more ape-like than *Homo erectus*. This species' ancestral neighbours, *Homo habilis*, so called because these beings used their hands, populated Africa between 2.1 and 1.5 million years ago.

From all these fossil records, we know that modern humans originated in, and then must have come out of Africa.

Fossil records on the island of Flores in eastern Indonesia threw light on the existence of yet another species of ancient human, *Homo floresiensis*. Remains show that this species was much shorter than we are today, around only 1.1m tall, and that the species' physical structure was likely to be much more like that of archaic humans and apes than of modern-day humans.

Your inner Neanderthal

Interestingly, sub-Saharan Africa has provided no evidence of the existence of Neanderthals there. This suggests that the Neanderthals' origins were different. Radiocarbon dating tells us that Neanderthals became extinct roughly 50,000 years ago. Potentially, then, there was opportunity for modern human and Neanderthal to have interacted. Studies of Neanderthal genomes reveal we did.

Using the four-population test, David Reich compared signature genomic markers of Neanderthals, chimps, present-day Africans and present-day Europeans. This exciting and groundbreaking work showed that 1–2% of the genomes of present-day Europeans contain markers of Neanderthal DNA, but that African genomes do not. The conclusion of this work was that Neanderthals and modern-day humans must not only have interacted, but that they must also have interbred, leaving their stamp on the genetic codes of the original Europeans with whom they crossed paths. The reverse was not seen, though, as studies looking for markers of modern-day humans in Neanderthal DNA found they were not present.

From mapping and comparing the genomes of these archaic humans with modern-day man, we now know that while one human species or subgroup was developing and existing in Africa, another had already made it into Europe. Not only that, we know that rather than these beings remaining distinct from one another, they interacted and bred.

▲ **DNA analysis shows that modern humans interbred with ancient humans before they became extinct.**

We don't know why Neanderthals eventually became extinct. Some hypothesise that perhaps modern-day humans persecuted them and out-competed them for the resources they needed to survive. Whatever the case, let the genomic record state that, in fact, Neanderthals do live on today – at least in genomic form.

Can we bring Neanderthals back to life?

In other words, can we use human DNA to clone Neanderthals? The simple answer is no, for the same reason that we cannot clone *Homo sapiens*. Even in theoretical terms, methods to perform human cloning are fraught with technological and genetic conundra so complex that they would be impossible to overcome. Ancient DNA extracted from bones and teeth is extremely limited, not least because it is incomplete. Because of these prohibitive technicalities, we will not be seeing a Neanderthal roaming this Earth again any time soon!

▲ **Neanderthal skull and reconstructured sculpture at an evolutionary theory exhibition, 2018. It is technologically very difficult to extract DNA from the fossilised remains of Neanderthals.**

Meet the Denisovans

The power of studying ancient DNA to discover ancient people really came into its own when sequenced ancient DNA enabled scientists to mathematically predict that there had been another ancient people aside from the Neanderthals. These people were closely related to, but different from the Neanderthals and they, too, had interbred with modern-day humans. The 2010 discovery in a cave in Denisova, in the Altai Mountains of Siberia, of a single bone and a single tooth provided the archaeological missing link. Sequencing the DNA from these artifacts revealed the predicted 'ghost' population, an ancient people now known as the Denisovans.

So, Denisovans and Neanderthals both interacted with modern-day humans – and yet, whereas we thrived, they became extinct. Analysis of ancient human DNA combined with ever more sophisticated mathematical modelling means scientists can infer from this growing body of DNA evidence how modern-day humans became established and how they migrated across the globe, out-competing archaic humans to become the dominant species we are today.

▼ The archaeological excavation site in the Denisova cave, Siberia, Russia.

Homo neanderthalensis or Homo sapiens neanderthalensis?

Whether or not Neanderthals belong to a separate species to modern-day humans (*Homo neanderthalensis*) or are part of a subgroup (*Homo sapiens neanderthalensis*) is the subject of heated debate. Strictly speaking, species are considered separate if there is no genetic interaction. The archaeological record had always suggested that this was actually the case, so *Homo neanderthalensis* should be their moniker. However, genome sequencing has now shown that there was a degree of genetic mixing: hello, *Homo sapiens neanderthalensis*!

The move into full speciation has occurred when there is no longer genetic admixture, which is thought to have been the case for the last 10,000 years before Neanderthal extinction, bringing us back round to the Neanderthals being considered a separate species – welcome back, *Homo neanderthalensis*. And so the debate continues...

The body beautiful

We can tell from the many Neanderthal skeletons available for study that Neanderthals were shorter and stockier than we are and that they had bigger brains. They lived in Europe in much colder conditions, so scientists think that their body shapes adapted to suit the environment. Messages stored in their DNA may yet provide enough information for us to work out a more exact hair, eye and skin colour.

With only a tiny finger bone and a large molar tooth to go on, we can't identify the physical appearance of the Denisovans. But as advances in DNA analysis come on in leaps and bounds, maybe one day the DNA record will shed light even on that.

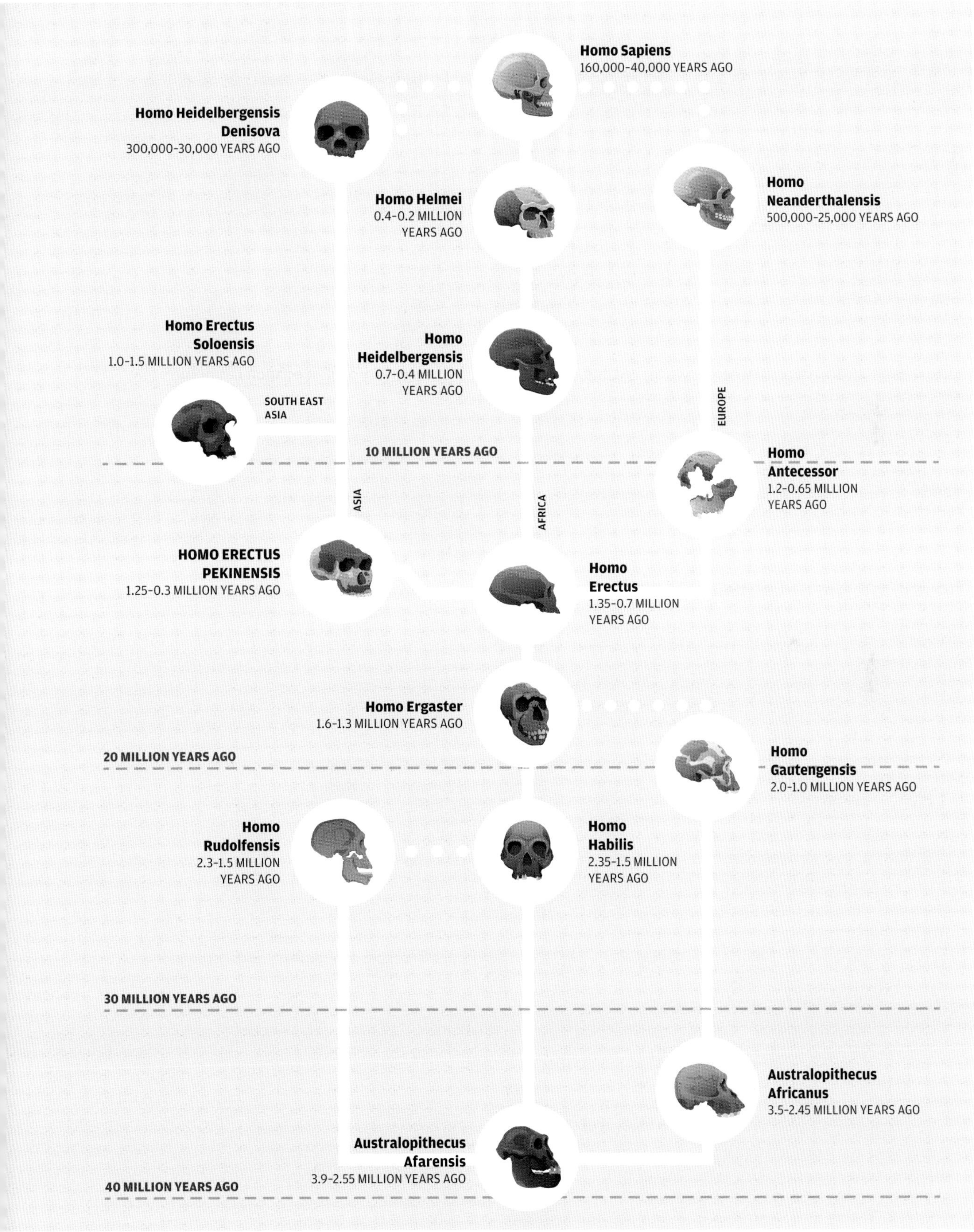

Homo Sapiens
160,000-40,000 YEARS AGO
Homo Heidelbergensis
Denisova
300,000-30,000 YEARS AGO
Homo Helmei
0.4-0.2 MILLION
YEARS AGO
Homo
Neanderthalensis
500,000-25,000 YEARS AGO
Homo Erectus
Soloensis
1.0-1.5 MILLION YEARS AGO
Homo
Heidelbergensis
0.7-0.4 MILLION
YEARS AGO
SOUTH EAST
ASIA
EUROPE
10 MILLION YEARS AGO
Homo
Antecessor
1.2-0.65 MILLION
YEARS AGO
ASIA
AFRICA
HOMO ERECTUS
PEKINENSIS
1.25-0.3 MILLION YEARS AGO
Homo
Erectus
1.35-0.7 MILLION
YEARS AGO
Homo Ergaster
1.6-1.3 MILLION YEARS AGO
20 MILLION YEARS AGO
Homo
Gautengensis
2.0-1.0 MILLION YEARS AGO
Homo
Rudolfensis
2.3-1.5 MILLION
YEARS AGO
Homo
Habilis
2.35-1.5 MILLION
YEARS AGO
30 MILLION YEARS AGO
Australopithecus
Africanus
3.5-2.45 MILLION YEARS AGO
Australopithecus
Afarensis
3.9-2.55 MILLION YEARS AGO
40 MILLION YEARS AGO

Home is where the DNA is

Advances in the field of ancient DNA technology have been rapid in recent years – scientists have now sequenced the genomes of over 1,000 ancient human specimens. Expanding the realms of possibilities to which we can apply this knowledge means ancient DNA analysis can build up a picture even of where different world populations originated and reveal historical human migration patterns to work out just how modern humankind made the world its home.

David Reich and his Harvard colleague Nick Patterson used combinations of genetic markers across 600,000 positions in the genetic code of individuals from 50 different modern-day populations. They compared and contrasted the markers to produce genetic signatures specific to each one. They used this information to draw up a map of our ancestors' migration patterns, based solely on the clues held within our genetic codes.

▲ **A map of Europe showing modern day borders – our ancient ancestors roamed freely, far and wide, leaving their DNA markers behind.**

The Ancient North Eurasian 'ghosts'

When Reich and Patterson arrived at studying the genetic signatures of northern Europeans, particularly the modern-day French people, there was evidence to suggest that the origin of Northern Europeans was somewhat confusingly halfway between Sardinian (commonly believed to be the first descendants of the farmers of Europe) and Native American.

When they applied their mathematical model to interpreting the genetic data, the scientists' calculations revealed a previously unrecognised population, people intermediate between early humans and modern-day man. They called them the Ancient North Eurasians (ANE), a 'ghost' population that could explain how northern Europeans could possibly be genetically linked to both Sardinians and Native Americans. ANEs, the maths inferred, were an extinct population that existed 15,00 years ago and contributed DNA to the genetic codes of both northern Europeans and to Native Americans.

Soon after this discovery, Danish geneticist Eske Willerslev, already credited with being the first person ever to sequence ancient DNA, sequenced the genome of Mal'ta boy. The boy, who lived 24,000 years ago, was discovered near Lake Baikal in central Siberia. Analysis of his genome showed him to be the missing link that Reich and Patterson had predicted. He was an ANE.

The Britons

Ancient DNA sequencing of samples from more than 200 skeletons from the Bell Beaker culture, which existed across Europe during the Bronze Age, showed the mixture of European populations in its make-up. One population came from the region of present-day Iberia and another, distinct group from central Europe with steppe ancestry, which spread into Britain 45,000 years ago. Analysis shows that the genetic background of these early settlers was similar to populations associated with the Bell Beaker culture in the Netherlands and replaced the minority population of farmers that must first have inhabited Britain and Ireland. So, we know that there was a wave of migration from central Europe, bringing steppe ancestry across the Channel and, along with it, the Bell Beaker culture.

Hunter-gatherers and pastoralist farmers

Extensive genetic mapping of migration across Europe first came in the form of mitochondrial DNA analysis, followed by genomic sequencing. Archaeological records already suggest that organised farming and domestication of the animals needed for labour was thought first to have begun in Anatolia, the most westerly projection of Asia. Constituting most of modern-day Turkey, this region was populated by migrants from the Eurasian Steppe, the grass plains extending from Hungary in the west to Mongolia in the east.

In 2009, analysis of ancient DNA from hunter-gatherers and from early farmers showed that these two have different mitochondrial DNA, indicating they have a different origin. When genome sequencing was brought into the mix in 2012, it became possible to sequence on a much greater scale. Scientists sequenced 200,000 markers in hunter-gatherers and farmers who lived in and around Sweden 8,000 years ago. These farmers were found not to be genetically like modern-day Swedes, but more like Sardinians. The hunter-gatherers did not resemble modern-day Swedes either, prompting further studies. David Reich's work on the subject points to all early European farmers deriving almost all of their ancestry from the original Anatolian farmers. Then, hunter-gatherers migrated from other areas and their DNA was incorporated into the farmers when the populations interacted.

▲ The grass plains of the Eurasian steppe extend from Hungary in the west to Mongolia in the east.

The Corded Ware, Yamnaya and Botai

Around 4,500 years ago, a new population emerged, the Corded Ware culture. Named after the cord-like impressions on the pottery associated with these people, genetically they derived their DNA from the Bronze-Age pastoralists, the Yamnaya people, hunter-gatherers from Eastern Europe and from the Caucasus. Migrating west from the grasslands of the Eurasian steppe, the Yamnaya may have been the people who introduced ANE ancestry into Europe.

Domesticating and taming horses to help people move from place to place made further migration across the steppe even more successful. Until recently, it was considered that the Yamnaya taught their pastoralist neighbours the Botai, originating from the Central Asian steppes of modern-day Kazakhstan and famed in archaeology for domesticating horses, how to ride. However, analysis of Yamnaya and Botai ancient DNA in 2018 showed no mixing at all, suggesting the two peoples never interacted. In which case, the Botai must have taught themselves to ride entirely independently of the Yamnaya people.

▲ The decorative cord-like pattern on the pottery gives the Corded Ware culture its name.

Indian subcontinent

Continuing to trace genetic migration eastwards confirms the conclusions of linguistic and archaeological studies that steppe migration led to India. To understand where people living on the subcontinent originated, we needed to look to mitochondrial DNA testing once again. The DNA sequence was unique to the whole subcontinent, so India's ancestors must have been isolated there for a long time without any genetic mixing. However, when scientists consulted the Y-chromosome, the results showed that there was in fact close relatedness to West Eurasians, suggesting that there had been mixing after all. Genome studies were required to settle the argument.

Scientists therefore turned to the Indian archipelago of the Andaman Islands, situated in the Bay of Bengal, to hunt for further clues to resolve the confusion once and for all. For thousands of years, since around the Stone Age, genetic isolation was guaranteed on Sentinel Island, one of the 300 Andaman islands that make up the archipelago, making it the obvious place to start when working out a pure genetic background with no mixing from outside.

Meanwhile, using mitochondrial and Y-chromosome DNA, this time combined with SNP marker analysis across the genome, populations from 25 different groups across India were tested with the idea of calculating their degree of relatedness based on ancestry. The results showed genetic variation across the whole population that was midway between European and East Asian, from a whole mixture of ancient populations: European, central Asian and Near Easterners (the West Eurasians) and South Asian (defined as related distantly to East Asians, which was closest to Chinese, but from a lineage that split 10,000 years ago). In fact, the people from the genetic isolate of the Andamans were found to be an isolated population of this South Asian branch containing no West Eurasian ancestry at all. So, DNA shows that the population of India, made up of 1.3 billion people, is a mixture of Ancestral North Indians (originally derived from the West Eurasians) and Ancestral South Indians (South Asians not related to any present-day population from outside India).

▼ Ancient DNA analysis reveals the origins of people living in the Indian subcontinent.

▲ **Analysis shows that 3–6% of the genome in indigenous Australians is Denisovan in origin.**

Chinese and Australians take the four-population test

Technological advances in ancient DNA analysis, developed in Europe and the USA, have not yet been fully realised in China, with only 5% of studies published carried out in South Asia. This, along with restrictions on exporting samples out of China, mean that little is known about the ancestral origins of this vast populous.

In Australia, there is archaeological evidence of human activity dated 47,000 years ago, as long ago as when the record shows modern humans were in Europe. One theory is that people migrated out of Africa and around the coast of the Indian Ocean, leaving descendants along the way, including in Australia, by the Southern Route hypothesis.

In 2011, Eske Willerslev conducted a four-population test on Europeans, East Asians, Chinese and Indigenous Australians. First, it showed greater ancestral commonality between Europeans and East Asians compared with the Australians, in keeping with the Southern Route hypothesis. David Reich's interpretation, though, raises the possibility that this similarity actually comes from Denisovan mixing. In fact, 3–6% of the indigenous Australian genome is Denisovan. Taking this into account, he proposed a model based on genetic data whereby Chinese and Australians originate from the same population. Their ancestors separated earlier than Europeans and interbred with Denisovans at a point in time following the mixing of Neanderthals with non-African people. Mathematical modelling estimated this event to have occurred around 54,000–49,000 years ago. This timing is certainly in keeping with the archaeological record.

Africans

In spite of modern humans evolving out of Africa, until recently scientists have sequenced only a handful of African genomes, dating from 5,000 to 15,000 years old. This is because the climate conditions in Africa prevent adequate preservation of the DNA. A groundbreaking discovery showing that we can more successfully obtain DNA from a certain part of the skull called the petrous bone (in the base of the skull) is starting to change this. Now, several projects seeking to analyse ancient DNA from across Africa are underway. The results of these projects will surely reveal to us the secret ancestral activities taking place in this vast continent while migration out of Africa was taking hold in the rest of the world.

Already, scientists have proposed that ghost populations are the reason we have discovered genetic ancestry markers in Ethiopia, Tanzania and Kenya. It's possible that these people even contributed to the out-of-African migration into Europe and the Middle East. With opportunities to access samples across Africa, and technological advances increasingly making analysis possible, the ancient DNA-mapping revolution that has transformed our understanding of migration. Ancestral history will finally bring knowledge of what happened in Africa, where humankind began.

▼ **The San people (or Saan), are the oldest inhabitants of Southern Africa.**

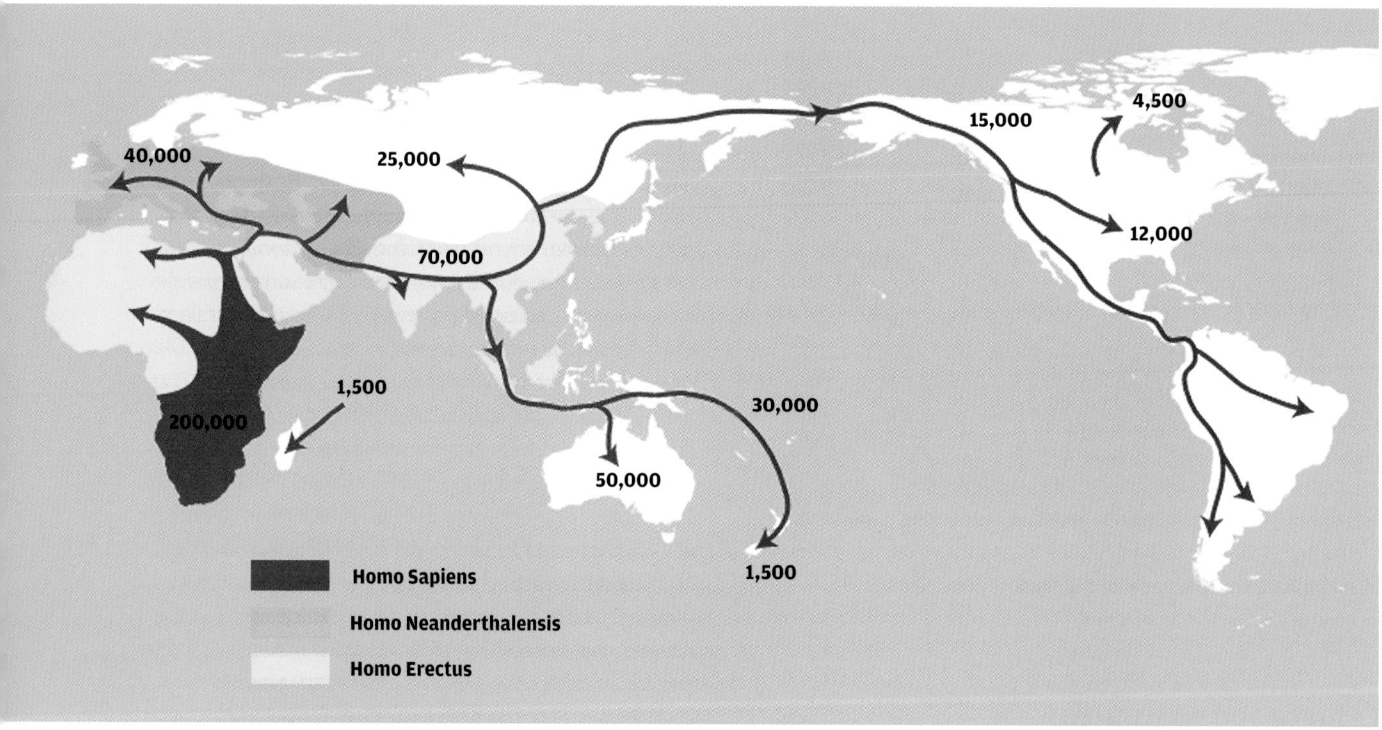

▲ **This map indicates human migration across the World showing the overriding success of *Homo sapiens*.**

Populating the Pacific Islands

Studying ancient DNA from the Pacific Island of Vanuatu tells us that the island was relatively recently populated, in at least two waves. First, people of the Lapita culture migrated in outrigger vessels from Taiwan, arriving 3,000 years ago. From here they populated other Pacific Islands. But people living there today have at least 25% Papuan ancestry. DNA analysis also revealed that this influx occurred 2,6000 years ago and was mainly male in origin. Papuan ancestral influence on other Pacific islands, such as French Polynesia and the Cook Islands, represented another wave, according to the DNA record, occurring 1,500 years after the first Lapita migration.

Modern-day South East Asians

In 2018, scientists analysed 26 ancient DNA samples from 8,000-year-old skeletons. This was no mean feat given that they were extracted from remains found in South East Asia, Malaysia, Thailand, the Philippines, Vietnam, Indonesia, Laos and Japan, where the warm weather conditions cause the DNA to degrade much more rapidly than it does in cooler climates. Analysing the DNA and comparing the genetic signatures to modern-day South East Asians revealed valuable information about their ancestral origin.

Archaeological records show that there were Hòabìnhian hunter-gatherers in the region 44,000 years ago, people so called after the stone artifacts discovered there and dating back to 20–10,000 years BCE. Before the arrival of ancient DNA technology, it was considered that these hunter-gatherers developed their own agricultural practices or, in the two-model theory, rice farmers from China migrated, bringing their skills to the Hòabìnhian.

The results show that Hòabìnhian hunter-gatherers were indeed the first inhabitants of South East Asia and they were joined by farmers from south China about 5,000–4,000 years ago. Many further waves of East Asian migration then came along, too. This not only supports both theories of early habitation, but also reveals once again that the reality is often a complex picture of multiple influxes of different populations.

Native Americans

Late in 2018, both David Reich's group and that of Eske Willerslev published work carried out independently analysing 64 ancient genomes from North and South America, dating from 10,900 to 5,000 years ago. The DNA came from samples originating from Alaska in the far north to Patagonia in the far south. The results show that Native Americans are genetically distinct. As to how they got to where they did, perhaps the genetic code can help.

Thirteen thousand years ago, Native American ancestors negotiated the Bering Strait between Russia and Alaska, when a temporary land bridge – called Beringia – formed just below the Arctic Circle. The DNA record tells us that these adventurous people split from Siberians and East Asians 25,000 years ago; and 15,000 years ago had links with the 'ghost' population of the Ancient North Eurasians (ANE). Having crossed the land bridge, some people stayed close to Beringia and others ventured farther south. These southward-travelling migrants then split into two groups: the Southern Native Americans and the Northern Native Americans.

The remains of a child, known as the Anzick child, who lived 12,700 years ago was discovered in Montana, USA. The Anzick child was part of the Clovis culture of mammoth hunters. DNA markers from this ancient Native American were compared with those of a 10,7000-year-old Spirit Cave individual found in Montana, and 10,000-year-old samples from Belize, Brazil and Chile and showed that they too had Anzick ancestry. This suggests rapid migration. Anzick ancestry has also been traced to South America from 9,000 years ago.

Mixing 6,000 years ago of DNA that we can link to the present-day indigenous Mixe people from Oaxaca in Mexico deeper into South America and 1,000 years ago into North America is also evident from the DNA record. But the full story held with the ancient DNA record is not yet complete.

Respectful burials

Detecting genetic markers that indicate Native American ancestry is also helping accurate identification of remains that belong to indigenous people once persecuted by European settlers. This, in turn, is assisting the repatriation of remains to their descendants. Before a flurry of 2018 studies, scientists had sequenced only six indigenous Native American genomes, but now – in large part because of the call for respectful burial – there is an upturn in interest in the secrets held in the DNA of these ancient remains.

The discovery of genetic signatures of Australasian ancestry found in individuals in Brazil, from people living today in the Amazon and an ancient sample from 10,400 years ago in Lagoa Santa, are so far unexplained.

▼ DNA from people in South America 6,000 years ago can be found in the indigenous Mixe people still living in Mexico today.

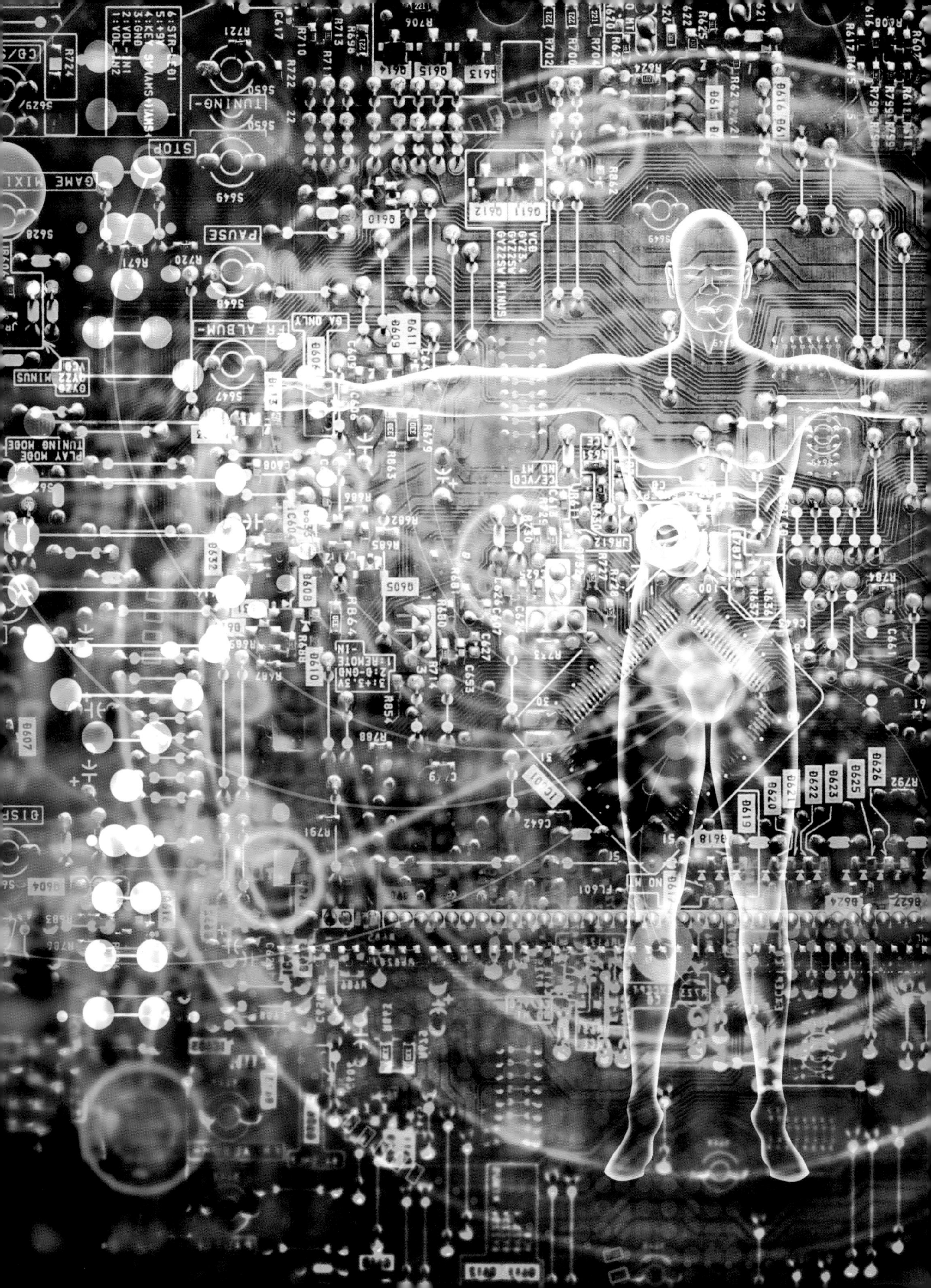

CHAPTER 4

Genes and evolution

Evolution is the process that describes how species have changed physically over millions of years, and we can link all living things back to common origins. The story of human evolution begins with the split from our primate ancestors. It is an account of the emergence and then extinction of our ancient human predecessors. And it refers to how we, modern-day humans, have undergone many adaptations to make us better equipped for living in diverse habitats. It even extends to the development of uniquely human traits, ones that have set us apart from other primates and mammals, both in physical form and cognitive ability.

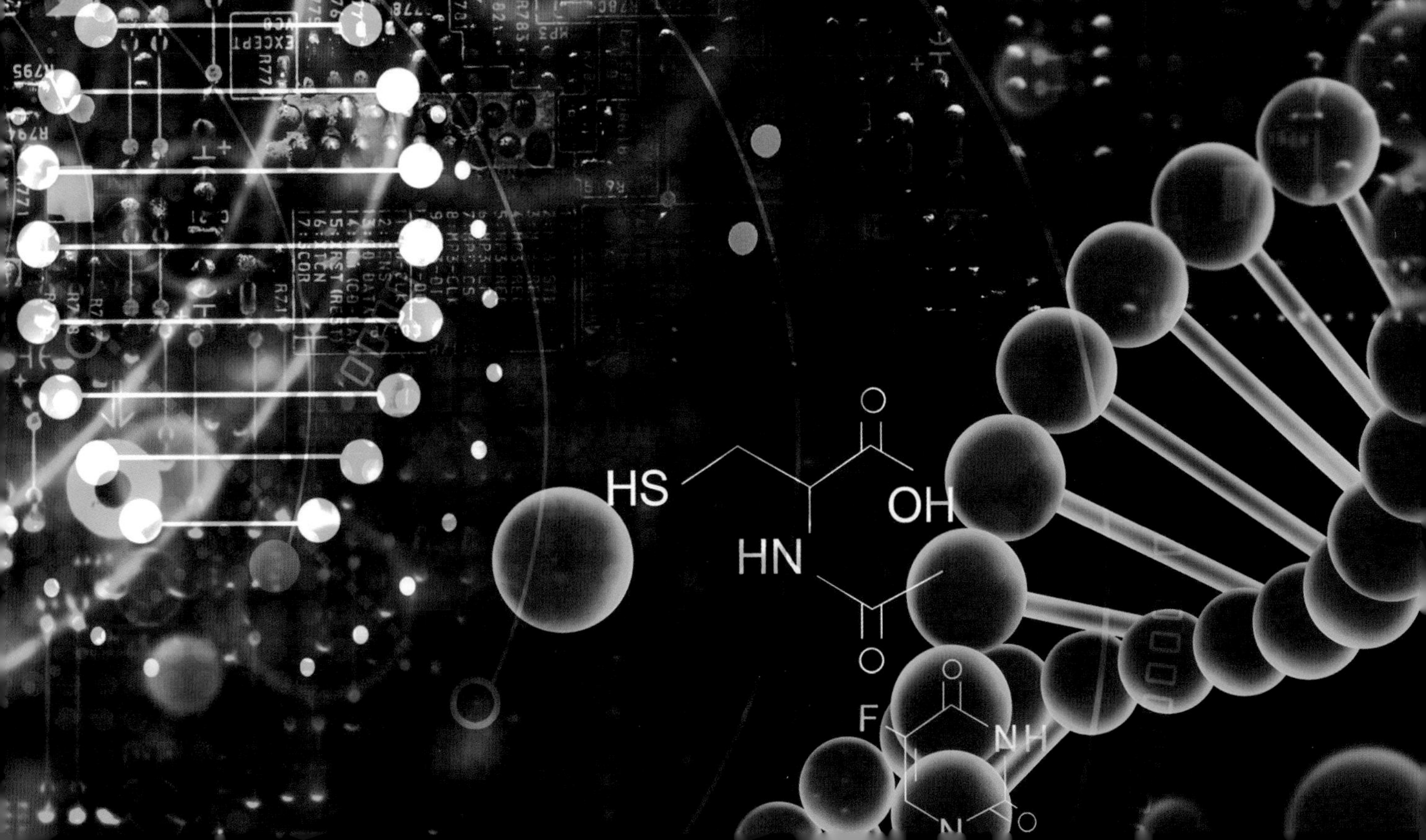

The history of evolution

The series of adaptations and changes – responses to environmental conditions – that marks out the timeline of human evolution took place over a prolonged period of time. Each adaptation was dependent upon randomly introduced changes in the genetic code (historically knowns as 'mutations'). These alterations affect proteins in only subtle ways, but when the subtle effects combine, together they make a big difference to biological function.

Where the shift in biological function gives one species an advantage over another, the body locks on to it, selecting it so that the species can thrive. Humankind has experienced an infinite number of these changes, which have accumulated over millions of years, each having a tiny effect in isolation, but together having the potential to create new species from old ones. We know all this because of incredible, collaborative work conducted hundreds of years ago.

Jean-Baptiste Lamarck (1744–1829)

In many ways French naturalist Jean-Baptiste Lamarck got the ball rolling. He believed that one generation could pass on to the next generation the physical features they had *acquired* during their lifetime in order to help the survival of their species – not only in their own generation, but also in the next. Taking the giraffe's neck as an example, he theorised that certain physical traits that are really useful (the long neck can, for example, reach leaves on higher branches), become more emphasised. Passing on these acquired physical characteristics from generation to generation means they become a much more prominent feature – and this is why, said Lamarck, giraffes have long necks.

While his early contribution to evolution was correct in some ways, his assumption about how a species transmits certain traits from generation to generation was not. The transmitted features are, in fact, the result of the accumulation of several genetic variants, each with a small effect, built up over many generations to create a more significant effect that beneficially adapts the species to its environment.

▲ **Jean-Baptiste Lamarck**

Charles Lyell (1797–1875)

A Scottish geologist, Charles Lyell took off on various exploratory expeditions to the island of Sicily, off the coast of Italy. Here, Lyell observed in great detail that Sicily's volcanic landscape had influenced the physical characteristics of the animals living in the region. He suggested that the species living there had originally migrated from Africa and Europe, but had

▲ **Charles Lyell**

A collaboration of minds

While Darwin was mourning the recent loss of his infant son, also called Charles, to scarlet fever, his friends and associates, Charles Lyell and the naturalist Joseph Hooker (1817–1911), warned him of the possibility that Wallace might beat him to an announcement. Having been shocked to learn that someone else had described his life's work, Darwin was all too happy for Lyell and Hooker to take care of the potential unseating on his behalf. They decided to submit Wallace's manuscript along with Darwin's theory of natural selection to the Linnean Society of London as a joint publication. Wallace was delighted and felt, ultimately, that it was his prompting of Darwin that pushed him to get on and write his *On the Origin of Species*. It was also Wallace who coined the term 'Darwinism'.

developed certain distinctive features that enabled them to adapt to their new environment. His work was a key influence for perhaps the most famous name in evolution history – Charles Darwin.

▲ **Charles Darwin**

Charles Darwin (1809-1882)

English naturalist Charles Darwin is famously credited with being the first to describe the theory of 'natural selection' as a major driving force of evolution. His book *On the Origin of Species by Means of Natural Selection*, published in 1859, sold out in just one day. Darwin's notion that we are all descended from a common ancestor (and not by intelligent design) was controversial, but he had developed it after a number of expeditions during which he studied the natural world. Voyages on HMS *Beagle* took him to the Galapagos Islands in the Pacific, where he painstakingly recorded detailed observations of the local finches and tortoises. He concluded these animals had physical features, suited to their environment, that meant they lived longer and had more offspring. He called this theory, the 'survival of the fittest'.

▲ **Alfred Russel Wallace**

Alfred Russel Wallace (1823-1913)

When Darwin got home from his voyages of discovery, he spent so much of his time musing over his findings that another British naturalist, Alfred Russel Wallace, almost beat him to publishing the notions of natural selection and survival of the fittest. Wallace studied natural history in the Far East and independently came up with the same conclusions as Darwin. Wallace shared his ideas with Darwin, whom he admired – and it was Darwin's wake-up call.

The evolution of evolutionary theory

Since Darwin's revelations, naturalists, archaeologists, anthropologists, geologists and biologists have all played a part in finding evidence all around us to support Darwin's theories of evolution and natural selection. Undeniably, though, in the modern world DNA takes centre stage as the primary evidence to unravel the secrets of evolution. Clues held within DNA lie at the heart of Darwin's theory and geneticists now have a permanent position in the team of evolutionary detectives.

The fundamental mechanisms that make evolution work are deeply rooted in observing the different genetic forces that make some populations successful and others not. Ever since Mendel studied pea plants in such exquisite detail, heredity and population genetics have been in the toolbox, ready to fill the gaps in understanding how evolution happens. But until the time was right for DNA analysis to take on a role, other sources of evidence – namely the fossil record and comparative anatomy – provided key information.

▼ It took 20 years to excavate this 3.6 million year old fossilised *Australopithecus* skull from rocks in South Africa.

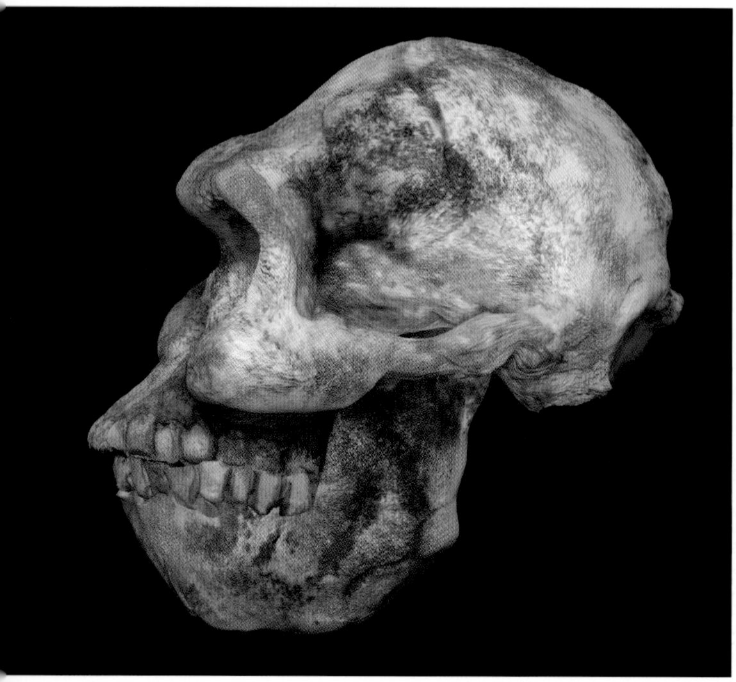

Evidence of evolution

Darwin explored both the differences and the similarities between species in his theory of evolution. He suggested that a new species, defined as one that can no longer reproduce with any other but its own, would arise when the cumulative effects of adaptation over many generations led to physical modifications that separate the old species from the new. Observing the subtle changes that occur over time brings evolution to life. Fossils provide one way for us to gain insight, and so does studying the bones of mammals, such as a whale and dog, that might seem to be unrelated but that actually have extraordinarily similar physical structures.

The fossil record

An historical treasure trove, the fossil record reveals information about species, including extinct ancient humans, that would otherwise be lost. In the case of humans, it gives us, for example, precious insight into what our ancient ancestors – hominins – might have looked like: fossilisation preserves body tissues, such as bones and teeth, that in life are hardened through mineralisation.

Most spectacularly, the discovery of Lucy and Ardi, skeletons dating back 3 and 4 million years, provides us with an opportunity to uncover the changes that must have taken place in early humankind both between these two examples (with an age gap of 1 million years) and between them and us (modern humankind), as well as between them and other specimens pulled from the fossil record. Such comparisons take us nicely on the way to painting a picture of our own physical evolution as humans.

Comparative anatomy

Literally meaning to compare the physical structure of one species with another, comparative anatomy identifies the features the two species have in common, linking them to a shared ancestor, and then studies the differences. These comparisons enable us to work out how the two species evolved in relation to their different environments.

To put this into practical terms, think about the forelimbs of mammals. Mammalian bones are in strikingly similar formation across species, even though one mammal might have the shape of a wing, another a flipper, another a cloven

Comparative arm anatomy

Comparing the anatomy of various different species shows that there are striking similarities, suggesting that in evolutionary terms they are descended from a common ancestor, with modifications built in to adapt to the environment they find themselves in.

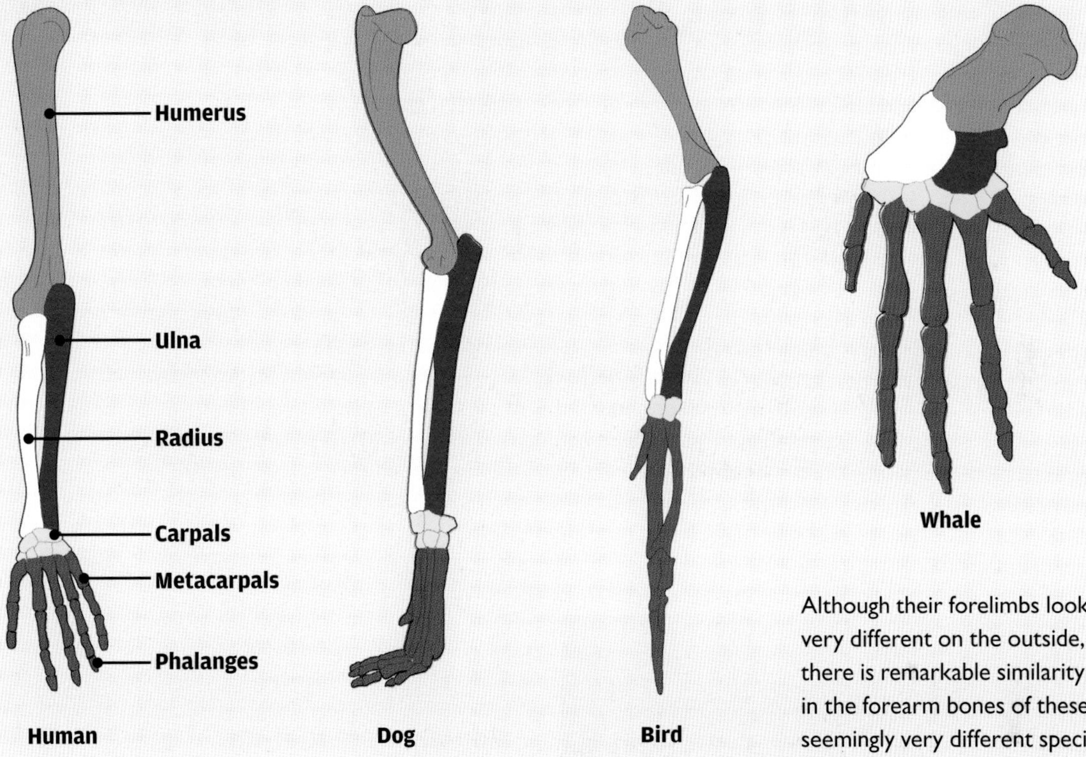

Although their forelimbs look very different on the outside, there is remarkable similarity in the forearm bones of these seemingly very different species.

hoof – and another an arm! These differences tell us that there exists a fundamental limb-patterning system that is shared across many different species. Evolution supposes that the explanation for this must be a common ancestor, and that, over a long period of time, different environmental forces must have influenced different species to evolve with various different adaptations.

DNA reveals all

Scientists have compared human mitochondrial and Y-chromosome DNA with that of other species in order to explore evolutionary patterns. Introducing genome analysis into the analytical mixture revealed that our genome is most similar to that of the chimpanzees. In fact, in genome terms, humans and chimpanzees are 99% identical. Looked at another way, chimp proteins have incredibly similar functions to our own. The chimpanzee, then, must be our most closely related primate ancestor. Genetics also tells us that we split from chimpanzees between 5 and 7 million years ago, calculated by estimating the time required for the variation in our genetic codes to have occurred through natural 'mutation'. (Now, of course, mutations are known as 'variations'.)

We can explain why different species have distinctive features in spite of sharing a common origin by looking at gene expression – a process that determines when, where and how much protein is made. Scientists are only now realising the full extent to which genetic diversity has played a role in evolution. Major developments in genetic technology, complemented by statistics and bioinformatics, have also revolutionised our understanding of evolution.

Genomic technology has enabled scientists to apply gene sequencing across the natural world, in ways that Darwin and Wallace would never even have imagined. We are now able to dig deep down into key regions of the genetic code across many different species, making comparisons and finding differences, this time gaining unprecedented access to the molecular record. This had led to the extraordinary discovery that there are areas within our genetic code that could hold the secrets to 'humanisation' – sections of our DNA that seem to make us human.

The role of genetics in evolution

Long before we could use genomic technology, scientists recognised that genetics played a pivotal role in driving evolutionary trends. They were also aware that there was more to it than 'just' natural selection. In fact, there are many other genetic mechanisms behind evolution.

The gene pool

The entire genetic catalogue of everyone who makes up a particular population, the gene pool is the collection of every single version of all the genes in that group. When one version of a gene, an allele, changes in frequency, so it becomes more or less common, and the shift can affect the way that population evolves. This can have a relatively powerful impact on the way physical features develop over a short period of time, a concept called microevolution.

Mutation

Random changes in the genetic code, which were once known as mutations and are now referred to as variations, occur constantly and have done for millions of years. Estimating how many variations have occurred in the genome of one species compared with another is a way to calculate how long ago it was that the two species might have had the same genetic code.

Genetic variation drives evolutionary change when it has a direct effect upon the way a gene works or when it changes the structure of the gene's protein, making it more or less effective. Variation is a main driver behind natural selection, as (in the reproductive stakes) new physical traits that are more advantageous to survival are favoured above all others. The overall effect can be that the favoured variation occurs more frequently in the gene in the pool, making it more likely to be passed on to subsequent generations and improving its chances of becoming a permanent feature.

Non-random mating

Non-random mating occurs when individuals preferentially choose a mate of a similar or of a different genetic background, influencing the gene pool. might follow from advantageous variations, a trait that might be beneficial, like being able to see further into the distance .

Gene flow

Introducing new versions of a gene to a population gene pool occurs, for example, when members of one population breed with members of an outside group. The results can alter the overall balance of traits within each population in shifts that may be advantageous or disadvantageous. We know from ancient DNA studies that modern humans mixed with Neanderthals, affecting the flow of certain genes that remain with us today (presumably because they were beneficial to both populations but originated in Neanderthals).

◀ **Genetic variation can result in physical variation.**

▶ **Human or beast, fish or insect, we are all here thanks to our DNA.**

Biodiversity

All living things are governed by the same DNA, but it is subtle changes in the genetic code that are responsible for the wealth of biodiversity on our planet

Genetic drift

Not every gene that occurs at a higher frequency in the gene pool is there because it's good for us, however. Sometimes one particular version of a gene can be more frequent just by chance. If you toss a coin 10 times, you have a 50% chance each time of achieving heads and a 50% chance of achieving tails. Overall, though, the 10 coin tosses may bring you 10 heads – or 10 tails, or any combination of heads and tails that adds up to 10. If, by random chance, one gene allele is passed on more often than others, its frequency in the gene pool therefore increases almost by accident. The effects of such genetic drift are more likely to appear in a smaller population group, simply because a wider, more diverse gene pool isn't available to dilute them.

Fixation

Fixation occurs when everyone in a population – all 100% of individuals – has a particular genetic variant on both copies of a particular gene, the one inherited from their mother and the one from their father. As this occurs in small populations, it is possible to study the fixed variants to see if their presence might have driven the development of certain traits that are considered uniquely human.

Natural selection

The best-known mechanism of evolution relies both on variability in the genetic code and heritability. Physical traits that make an individual better suited for the environment in which he or she lives are selected above others because they provide a useful adaptation that increases the chance of the species' survival.

The initial genetic variation that can change the function of a protein is usually both random and has only a small overall effect. But, in subsequent generations, there may be positive selection for that genetic change and its physical consequence, especially if it means better chances of survival. The more frequent the variant becomes, the more likely it is that the species will adopt it as a permanent fixture.

Once the species introduces further random genetic variation, the adopted physical trait can become even more specialised as an environmental adaptation. Darwin most famously described this concept as a major evolutionary influence in the beaks of the various species of finch living on the Galapagos Islands. Each species had a beak a different shape and size to the others, reflecting the need for the different finches to find food from different sources. Some Galapagos finches fed on nectar, for example, and their beaks

Bottlenecks and founders

The bottleneck effect occurs when there is much lower genetic diversity because of significantly reduced population size. This might happen when a natural disaster wipes out large numbers of an original population and the genetic codes of the survivors are not representative of the codes that were more commonly present in the whole.

The founder effect occurs when a small group splits off from a larger and therefore more genetically diverse population and establishes a new colony. The resulting gene pool might not reflect that of the original group – only that of the new colonisers.

Founder effects are often seen in medicine when rare diseases occur with much higher frequency in geographical pockets of a population than they would by chance. In effect, healthy 'carriers' of the gene causing the rare disease have a greater influence over the gene pool because, as a splinter group, it is so small. The carriers might be some of the found splinter group themselves, or explorers who arrive into the splinter group and inter-breed. A genetic blood disorder called variegate porphyria, which is much more common in South Africa than other parts of Africa or of the world, is a good example: scientists have traced the introduction of this harmful gene copy back to a Dutch couple who migrated to South Africa in 1688. The pathogenic variant was passed down to subsequent generations, so a high frequency of carriers built up in their descendants, greatly increasing the chance of the condition affecting the population.

▲ Isolated populations, like the first Dutch settlers in South Africa, are more likely to be influenced by genetic drift, fixation and natural selection.

Adaptive radiation

Adaptive radiation occurs when one ancestral species gives rise to a number of new forms through different adaptations of the original species. This process is best illustrated by Darwin's finches, birds isolated on the Galapagos Islands, derived from the same ancestor, but developing alternative physical features to adapt to different aspects of their environment.

▶ **Darwin's theory of natural selection came from his studies of finches on the Galapagos Islands.**

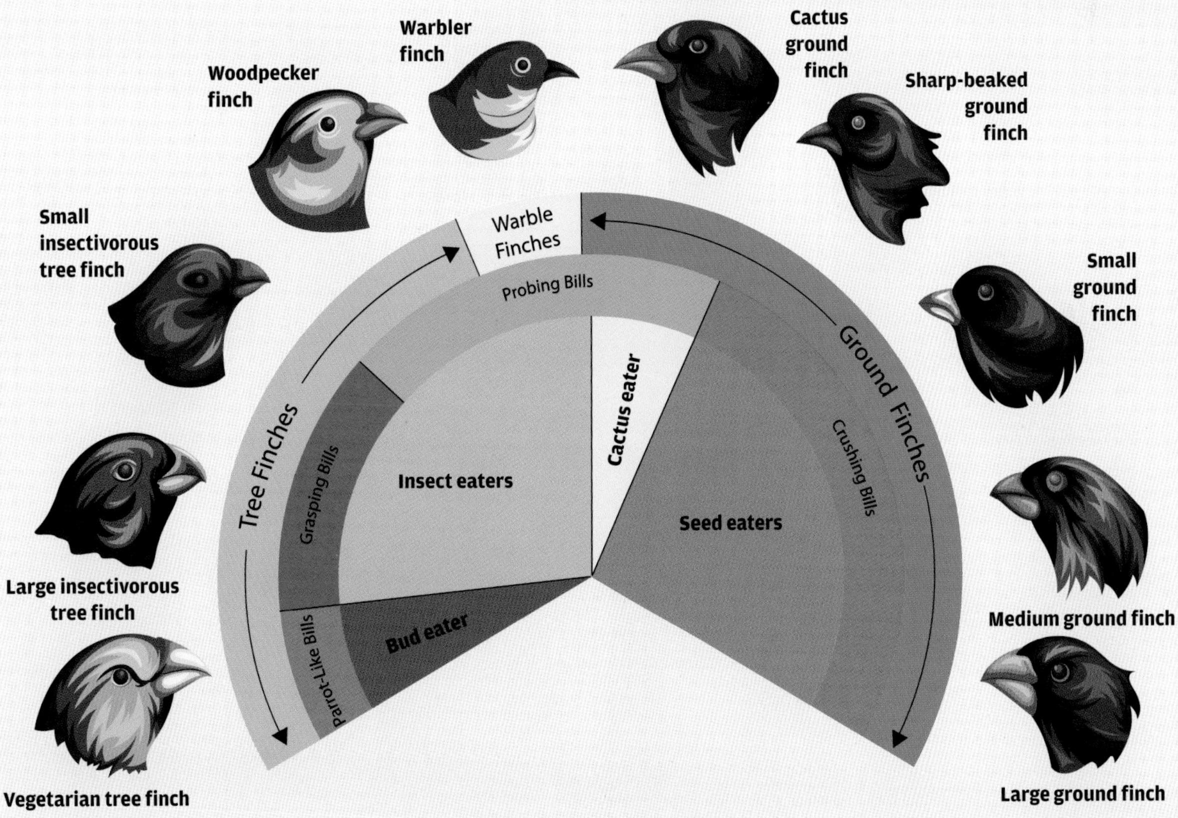

had adapted to become a different shape and size to those feeding on seeds, which were different again from those feeding on insects. Scientists from Uppsala University in Sweden sequenced the genomes of several species of finch and found a number of variants in a gene called ALX1, nicely illustrating Darwin's notion that genetic variation underlies the differences he observed.

The effect on human evolution of genetic variation that resulted in natural selection and the so-called 'survival of the fittest' is one of the main mechanisms that explains how our species adapted and evolved to create the immense diversity we see in it today.

Reproductive fitness

An individual's success at reproducing is known as his or her 'reproductive fitness'. 'Success' is a measure of the number of organisms the parent produces with the desired genetic make-up – their DNA legacy after they have gone. In genetic terms, this means the parents have left their genetic stamp on a population. If, over many subsequent generations, the stamp instills a physical trait that better adapts the offspring to the environment, then microevolution has taken place. In effect, fitness, the environment and natural selection work together to effect change.

In practice, no single advantageous genetic change has such a large effect as to drive microevolution on its own. Instead, a species needs multiple subtle changes involving several different genes to have a collective effect that drives evolution over thousands of years.

▼ A number of small-effect genetic changes can have a larger cumulative effect to drive evolutionary processes.

Polygenic traits

Some genetic determinants for a change can be polygenic. In other words, they can involve multiple genes. Each gene exerts just a small effect upon a particular physical characteristic, then together they are greater than the sum of their parts: creating something much more significant than each minor change suggests. Variation in the combination of gene variants from one individual to the next results in a whole spectrum of physical differences we see occurring normally in a population.

For example, height is a polygenic trait. Scientists have identified genetic variants in more than 700 genes as having an effect on height. Individually they might not have much of an effect at all, but collectively they add up to have a major influence. The outcome means that one population may exhibit a whole range of heights at any given age.

Think of the growth charts we use for babies and children, which give a centile range showing the average

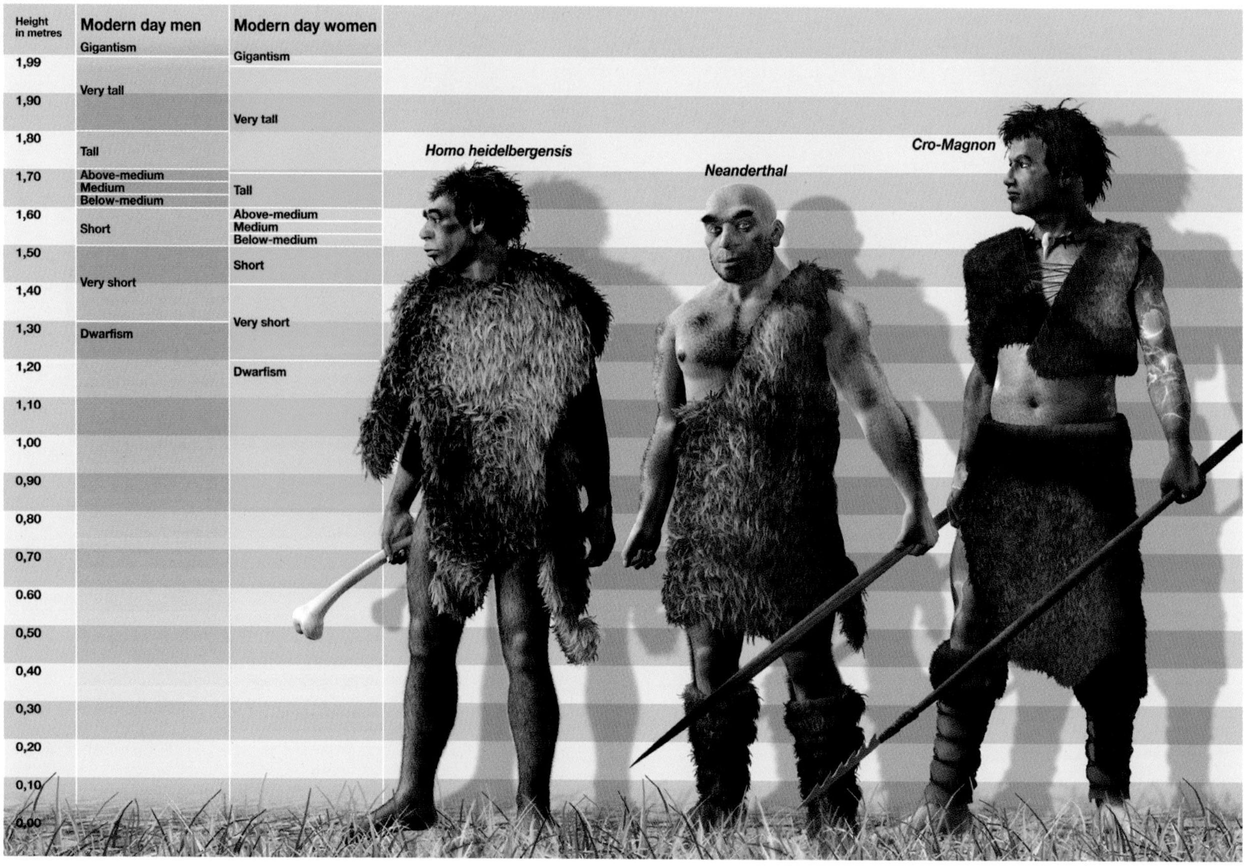

Different mechanisms of natural selection

Not all natural selection occurs in the same way – there are three different types: stabilising, directional and disruptive. All are selective forces on mainly polygenic traits, meaning that there can be both extremes of a trait in a population. The trait may either occur normally distributed over the whole range between the extremes, or occur at just the extremes. Which mechanism predominates depends on the environment driving the physical change.

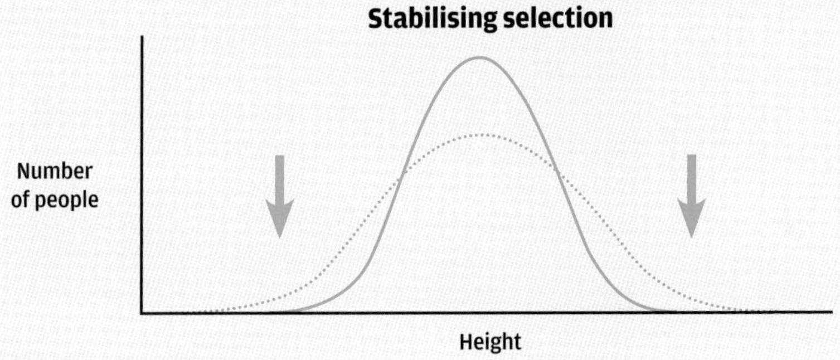

Stabilising selection is represented by the 'bell curve', when advantages within the prevailing environment are evident at all different heights, for example.

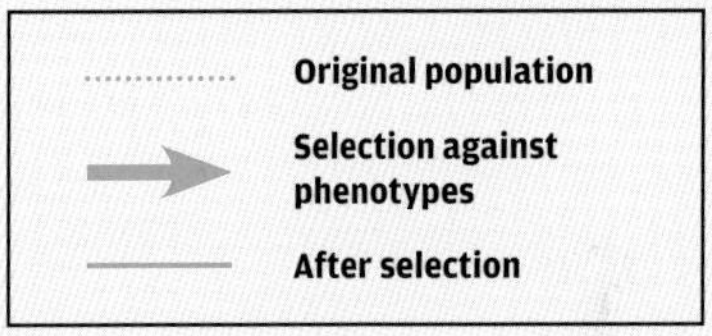

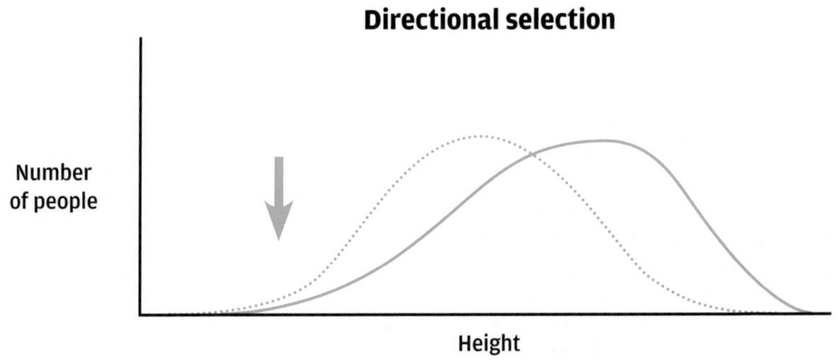

Directional selection is more likely to occur when one particular physical trait is more advantageous – for example, being very short.

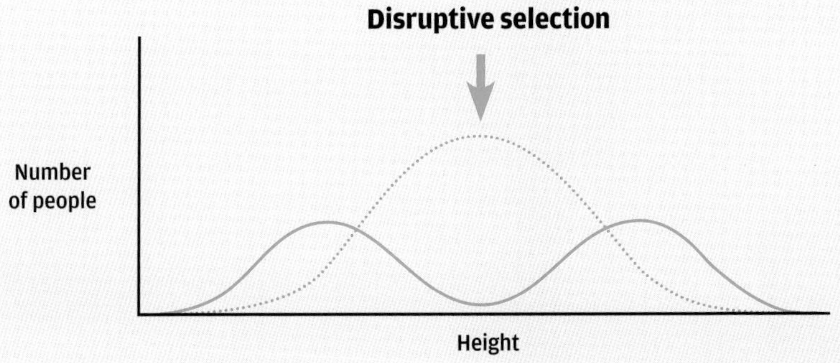

Disruptive selection occurs when extreme ends of the spectrum are represented – for example, very tall or very short – where it is the environment that dictates which form predominates.

and the upper and lower limits of expected height for girls and boys at a particular age. This is genetic variation displayed in graph form.

Alterations affecting the biological function of the genes affecting height can have a major influence. The fossil record tells us that a population of ancient hominins once lived on the island of Flores in the east of Indonesia. Standing at 1.0m (3'3") tall, the average height of *Homo floresiensis* was much shorter than ours today. Whether this shortness was the consequence of limited resources on a small island stunting growth, whether the population was genetically programmed to be this height (a polygenic phenomenon), or whether some other significant genetic influence was at play to cause restricted height is unknown.

Humanising humans

We know from comparing DNA that chimpanzees are our closest ancestral relatives and that our genetic codes are almost identical. Yet physically we are very different beasts. So what makes us human? Could the answers lie in the 1% of the genetic code where we are not the same?

Human accelerated regions

In 2006, Katherine Pollard, a biostatistician who was comparing genetic differences between tumours and healthy cells to find out what drives cancer, used the same scientific techniques to compare chimp and human genomes. Her idea was to study those sections where the codes are not the same and work out what their role is.

She initially found 202 of these areas of interest on the genetic code, places that she thought might be potentially responsible for 'humanisation' (what makes us human). She called these 202 sections 'human accelerated regions' (HAR), reflecting her theory that there might be hidden messages within these genomic locations that might have driven the process of variation that exists between us and our ancestral relatives, chimpanzees.

Unexpectedly, most of the HARs are not actually contained within genes themselves. They are found inside non-coding regions – that so-called junk DNA. Typically, the HARs are just over 200 base pairs long and are found close to genes that play a role in physical characteristics that distinguish us from the chimpanzees. They must exert some control over the way the genes work – for example, when and where their proteins are made – making them the so-called regulators of gene expression. And it is these differences in the way our specifically human proteins are produced that make us uniquely human.

▼ Ardi, the 4.4million year old skeleton of *Ardipithecus ramidus*, was partially bipedal.

Human traits

So what are the characteristics that make us this way? How do we actually compare to other primates?

Bipedalism

The fossil record tells us that Lucy – the 3-million-year-old skeleton from the early hominin species, *Australopithecus afarensis* – was already walking on two legs, but Ardi – the 4.4-million-year-old *Ardipithecus ramidus* skeleton – was only partially bipedal, more like the knuckle-walking chimpanzees. We can only speculate why walking on two legs is a good thing: it frees up our hands to use tools; it lets us see further into the distance, to identify threats and better habitats to explore; and it means we can pick berries that might otherwise be out of reach.

But being bipedal also comes with disadvantages. If you have ever wondered about the seeming injustice of the pain that giving birth in humans brings with it compared to other mammals, it is all down to the size and shape of the pelvis. The design of this bony girdle had to alter – becoming narrower and less efficiently orientated for childbirth – to let us walk upright. The shape of the spine had to change too, as did the base of the skull – all to accommodate the weight of the head, aligning us so that we could balance.

A number of HARs have been found linked with genes important in the development of the embryo at stages when the overall structure of the body is being determined. The genetic instructions that come into play during this process – called body patterning – are regulated in different ways from animal to animal, even though the underlying genes are either identical or extremely similar. This has led scientists to believe there may be a connection between HARs and uniquely human traits – such as walking on two legs – as a result of the way HARs influence gene expression during early development.

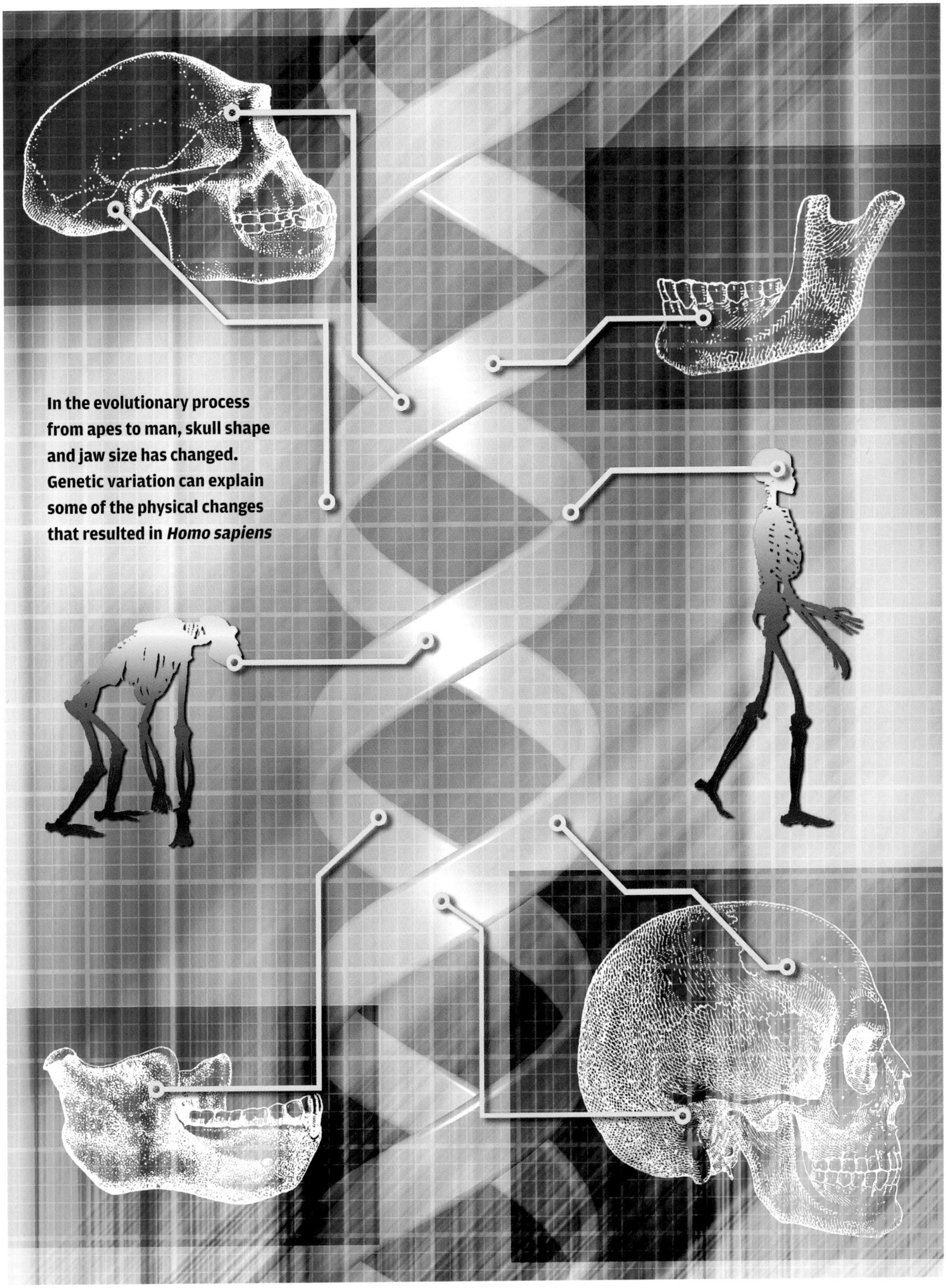

In the evolutionary process from apes to man, skull shape and jaw size has changed. Genetic variation can explain some of the physical changes that resulted in *Homo sapiens*

Brain size

Katherine Pollard identified an enhancer HAR called HARE5. It increases the production of the protein of a nearby gene called *Frizzled8*. The action of this gene is much greater in humans than in chimpanzees, making the brain larger – one of the key evolutionary differences in the physical structure between between the two.

It is likely, though, that there were a number of factors acting together over time that resulted in humans having larger brains, not just this HAR. Exactly how this key physical development came under the force of natural selection remains a mystery – but as it's about the very thing that most significantly sets us apart from the primates, let's hope we work it out sometime soon!

Face shape

Using the fossil record as a guide, we believe that the face shape of modern-day humans is different to our hominin ancestors, such as the Neanderthals. There is, a genetic basis to this, although we don't know whether the genetic changes were under the control of natural selection or the result of chance over thousands of years. We do know that our brows are less prominent, our foreheads more vertical and our noses shallower than our ancient predecessors'. Our jaws are a different shape, too – the response to changes in our diet perhaps, as over time we have created less demand on our jaws to chew: not only is our food softer that it was in prehistory, but our dexterity meant we developed tools to cut up our food before we put it in our mouths.

In 2016, researchers identified a handful of genes whose expression has changed over time in line with changes in our face shape. By comparing the workings of certain genes across different populations with distinctive differences in the face shape – such as the width of the nasal bridge – they identified variations that could be driving the differences. For example, markers within *RUNX2*, a bone-growth gene, were linked to the width of the bridge of the nose, whereas *GLI3* and *PAX1*, genes that control cartilage formation, where found to be important in the width of nostrils.

However, we do not have any strong evidence for how these features might have adapted us better for our specific environments. Certainly, other physical features are more obvious candidates for clear adaptation, such as skin colour and living at higher altitudes.

Skin colour

Genetic variation in certain genes results in different skin, hair and eye colour. For example, certain markers in the *MC1R* gene are more common in people with red hair and fair skin – features that are much more likely to be found in people from northern European backgrounds, as opposed to those from Africa. Why should skin colour become paler the further north we are?

Skin pigment protects us from very high levels of ultraviolet

▼ Researchers have identified a number of genes that influence the shape of our faces, including our skull, nose and chin. Using ancient DNA it will be possible to look at how these genes have changed during evolution, so we can go back and construct the facial features of our ancestors based on their DNA.

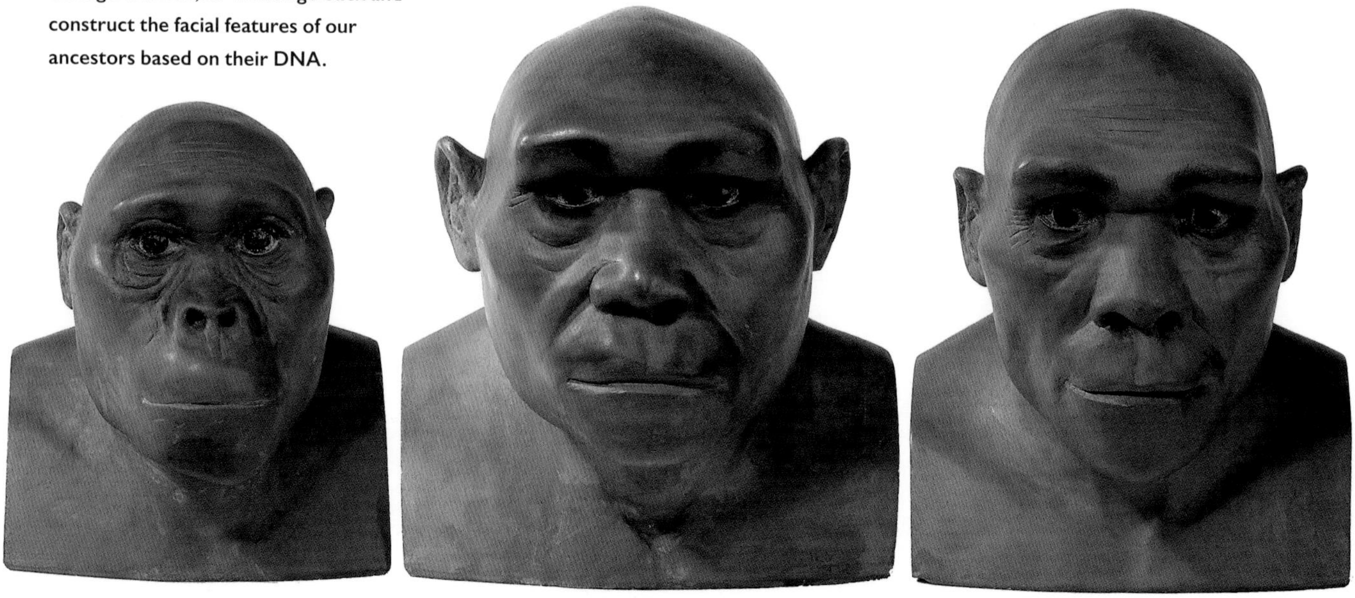

Australopithecus **Early Homo erectus** **Late Homo erectus**

(UV) radiation (from the sun), which is particularly strong around the equator. Not only that, though, the skin is a major source of vitamin D, the most biologically active form being produced under the skin following exposure to UV rays. Paler skin makes it easier for the UV rays to penetrate the skin. And then a further adaptation, to the vitamin-D receptor gene VDR, has made it easier for the body to synthesise vitamin D from the UV rays of the sun. This combination has meant that those living further away from the equator are efficiently adapted to their environment, in turn making more of the world habitable.

The power of speech

Children with damaging variants in the *FOXP2* gene have significantly impaired ability to speak, so this gene must be important for communication. The protein that the gene expresses is the nearly same in every species, but between chimpanzees and humans there is a very subtle change, involving just two of the amino acids (the building blocks of our proteins) – which must be the reason that we have the power of sophisticated speech and chimps do not. Interestingly, our Neanderthal relatives had this version of the gene as well, so it is tempting to think they too spoke to each other in something more than an incomprehensible grunt!

Communication

Our ability to speak along with our changing face shape may have revolutionised the way our hominin ancestors communicated. The more vertical shape of the forehead and the less heavy brow allows us to have more expressive faces than early man. As these adaptations occurred, they may have paved the way for more effective social interactions and cooperation between individuals, allowing *Homo sapiens* to evolve more rapidly as organised groups, protecting and nurturing each other, working together to get the best out of their environment and ensure survival.

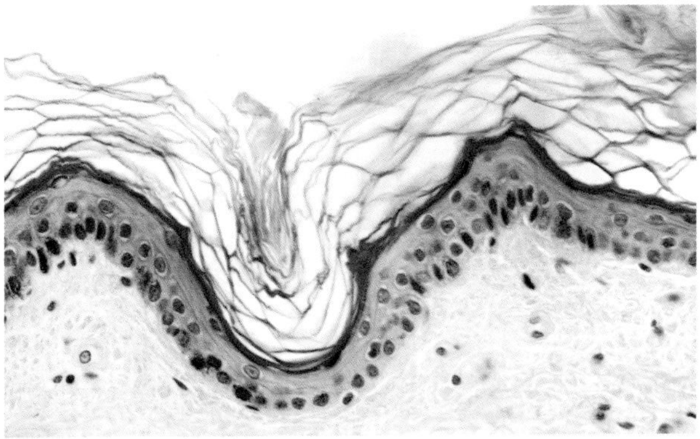

▶ Pigment in the skin is produced by cells that make melanin, seen here under the microscope as small brown dots in the epidermis.

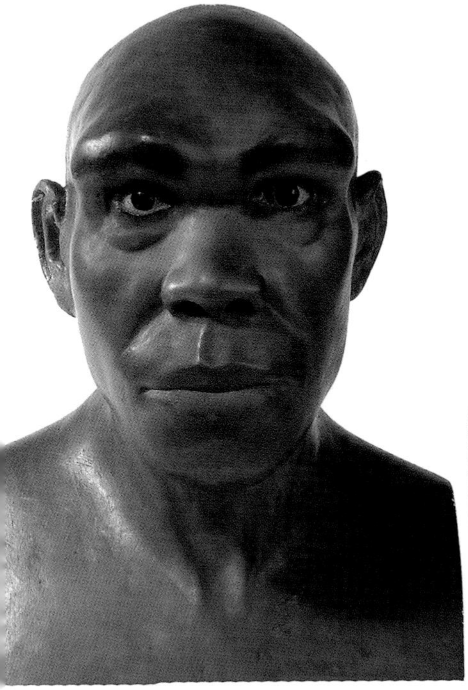

Homo heidelbergensis

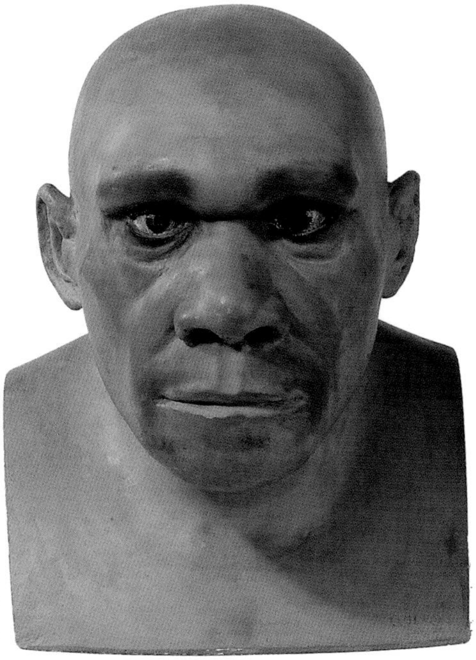

Homo neanderthalensis

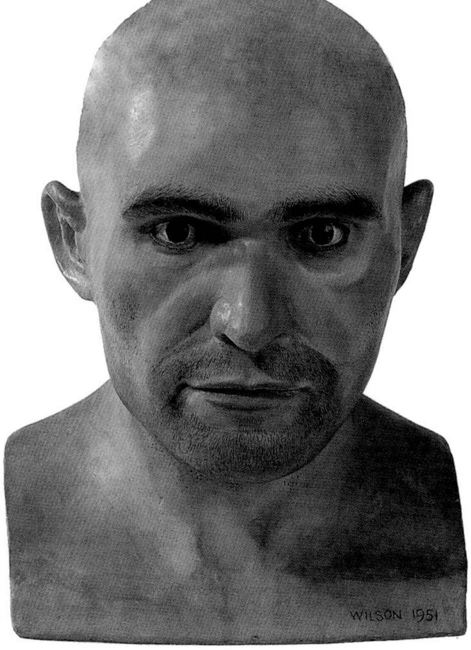

Early Homo sapiens

Survival of the fittest

A key principle of Darwin's theory of evolution, survival of the fittest acknowledges that certain physical characteristics are more likely to ensure a species' survival through better adaptation to the environment, come rain or shine.

Protection from disease

Being a carrier of a disease that in its full-blown form will cause major medical conditions (such as sickle cell anaemia) can help protect you from other illnesses (such as malaria). These effects are widely held to be the reason why there is such a high carrier frequency of some genetic conditions in certain populations. For example, 1 in 20 people of African origin carry the gene variation that causes sickle cell.

Another such example is the relatively high frequency of cystic fibrosis (CF) carriers in Caucasian people. CF is a severe genetic respiratory condition that causes progressive lung damage. Lung secretions in those with CF are thick and sticky, because the mucus that ordinarily protects the lung lining is abnormally salty – the result of an abnormally functioning chemical channel across the cell membrane.

Amazingly, CF carriers have some protection against cholera, an infection by the *Vibrio cholerae* bacterium that affects the intestine and, left untreated, can result in death. The cholera bacterium releases a toxin that attaches to the cells of the gut, altering the flow of salt across its membrane, resulting in watery diarrhoea and serious dehydration. The toxin functions in the opposite way to the abnormal salt channel in CF – which means that at times when cholera has been rife, CF carriers might have had a selective evolutionary advantage over non-carriers.

Lactase persistence

Infants drink milk and, for the first few months of life, it constitutes their whole diet. Milk contains lactose, a disaccharide sugar that the body has to break down into its constituent parts, glucose and galactose, in order to digest it. The trigger for this process is the protein enzyme lactase. The gene that produces lactase is programmed to switch off in later life – in other words, we stop producing it during adulthood. At least, that is the case for people of Asian descent, but Europeans have a genetic variant that keeps the gene actively making the milk-digesting

CFTR proteins

The cystic fibrosis transmembrane conductance regulator (CFTR) sits across the cell membrane, letting salts in and out of cells, left. Pathogenic variants in the gene that makes CFTR prevent the flow of salts, and very sticky mucus builds up in places like the lungs, resulting in cystic fibrosis (CF), right. But the in-between effect on CFTR in CF carriers is thought to provide some protection against the toxic effects of cholera infection.

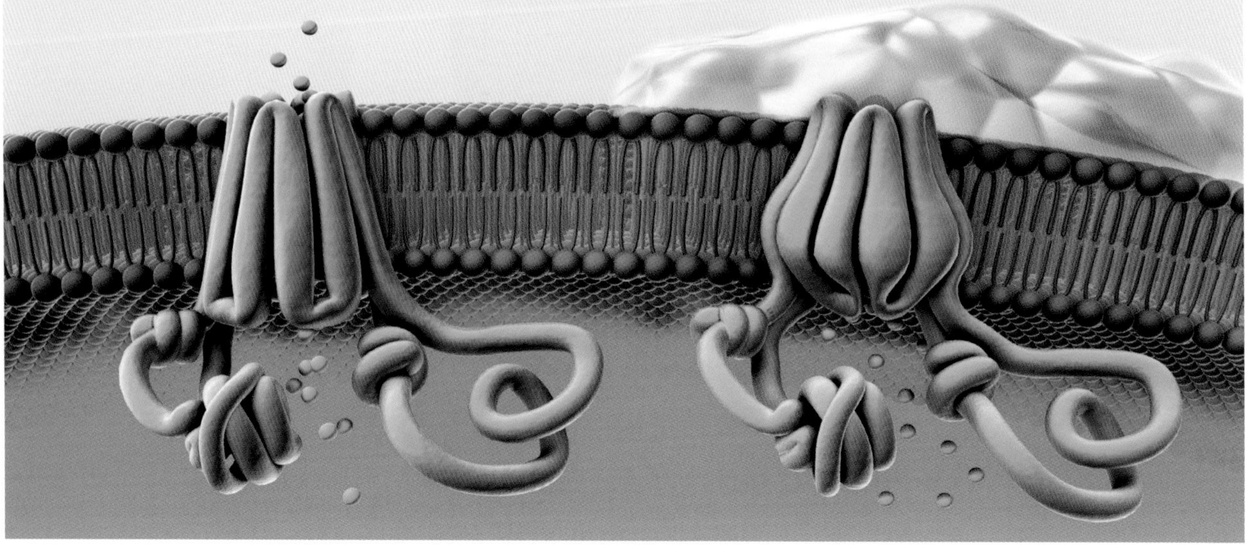

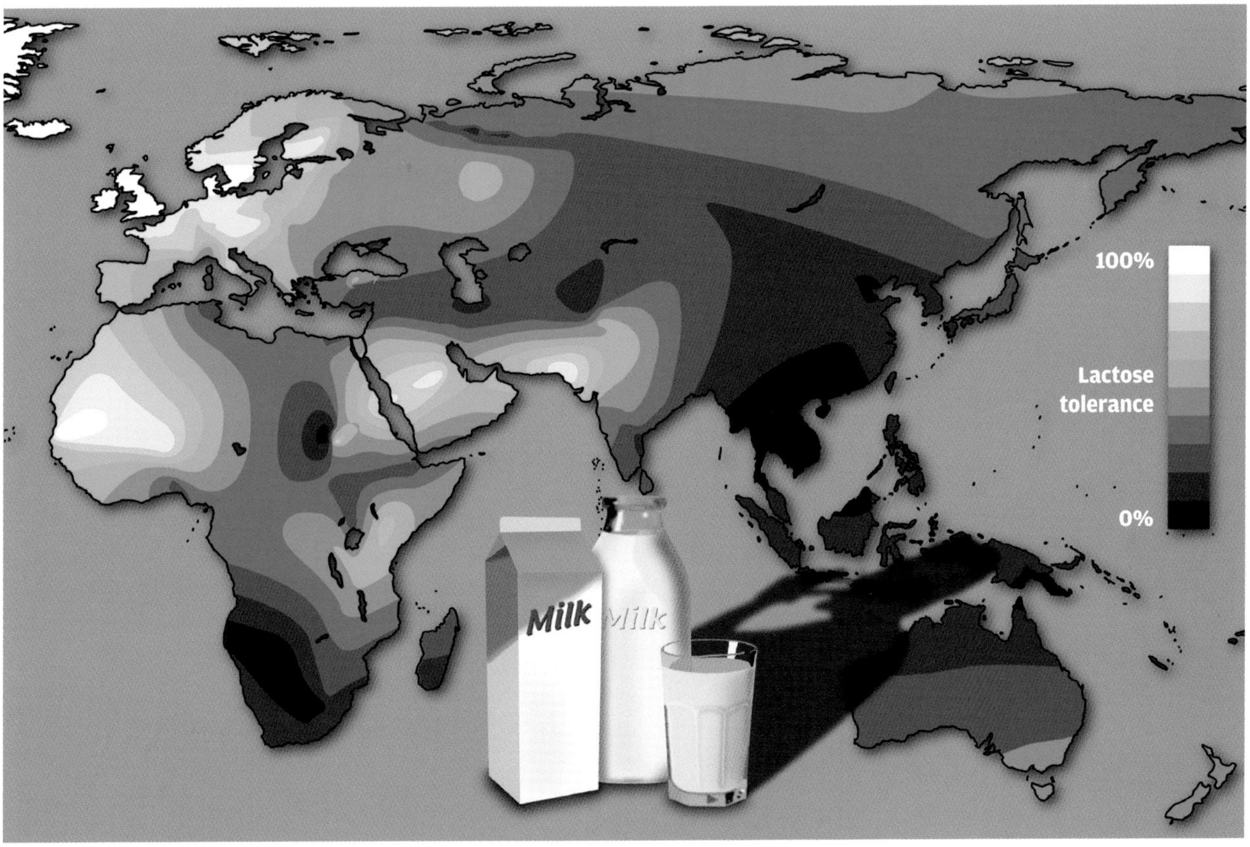

▲ **A map showing the pattern of lactose tolerance in populations across Eurasia, Africa and Australia.**

enzyme. There are a handful of variants found in African and Middle Eastern populations, too. Why?

In order to find an evolutionary explanation for the development of Neolithic farming practices 8,500 years ago, scientists are exploring possible geographical links with lactase persistence. Dairy farming and the ingestion of a lactose-rich diet will have necessitated the continued production of lactase in farming populations. So, building up lactose tolerance might just be another example of natural selection at work – a physical adaptation some factions of humankind have needed to exploit their specific environment.

A perfect match for life on Mars?

The combination of scientific disciplines – such as paleontology, anatomy, biology, genetics and archaeology – creates a powerful tool for mapping evolutionary trails in the quest to understand better who we are and where we came from. To have the opportunity today to bring scientific rigour to the theories proposed by these founders of evolutionary theory is astounding – a 21st-century gift to help fathom the human condition through genetic technology.

It will take all that we have in the field of genetic code breaking to demonstrate exactly which parts of our genomes drive evolution and how it was achieved in such a seemingly well-thought-out way. We may even be able to watch evolution in motion. Perhaps we could one day identify genetic factors that could equip us, in true natural-selection style, for life on Mars…

Always getting a bad hangover?

The body has specific proteins, the function of which is to clear alcohol and its toxic by-products from the body. These so-called alcohol dehydrogenases (ADH) are found in the liver – and the faster they work, the better we feel the morning after the night before.

The genes that produce ADHs, like all other genes, contain variants that affect efficiency. Slow metabolisers are not as efficient at clearing alcoholic waste as fast metabolisers, and will feel worse in the morning. Variations in these genes have also been linked with a predisposition to chronic alcoholism – although not in a predictable way just yet.

Darwin's tubercle and other vestigial traits

Ever wondered what the appendix is for and why it is so often the troublesome object of the surgeon's blade? Why do men have nipples? Why do wisdom teeth impact and have to be so brutally extracted? And what do we need sinuses for when all they seem to do is cause painful congestion? These and other traits are thought to be the legacy of our evolution from an earlier time when these were potentially useful parts of the body.

Our Denisovan and Neanderthal friends were known to have much larger molar teeth than us, presumably so they could easily grind plant-based food. Having more teeth at the back of the mouth would increase the surface area for

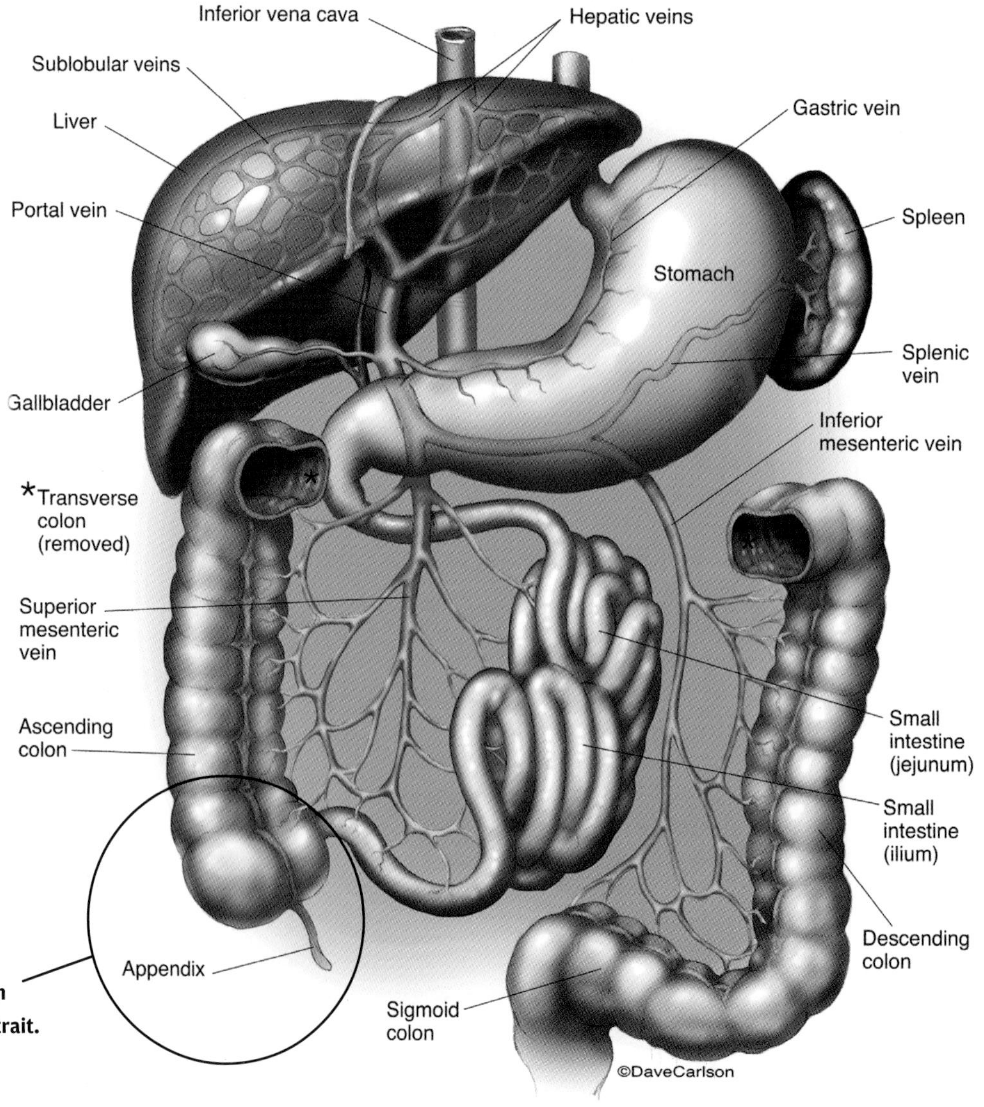

The appendix, circled, is viewed by many as an evolutionary vestigial trait.

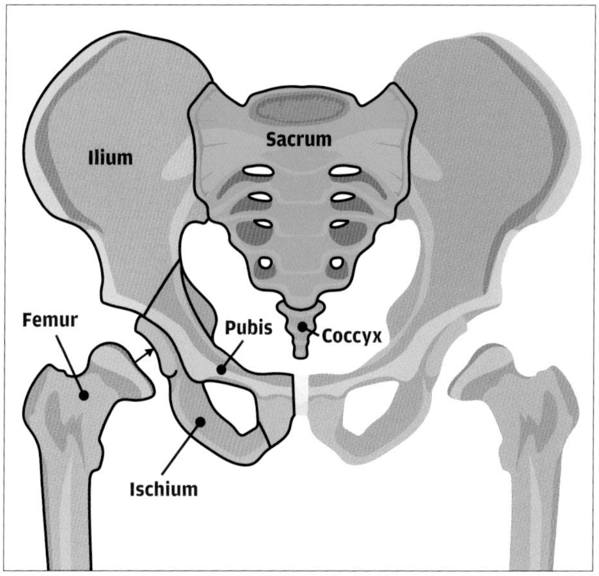

▲ **The coccyx, or tailbone, is an anatomical reminder of our primate past.**

▲ **A baby baboon with prominent points at the top of the ears. Humans have remnants of this trait, known as Darwin's tubercle.**

chewing in a similar way, but now we no longer need wisdom teeth to manage our softer, modern-day diet. Up to 35% of people never develop wisdom teeth, suggesting evolution is still at play and there is no selective disadvantage any more if we do not have them.

The appendix has mystified scientists for years and plagued the one in 20 of us who have had appendicitis. The only treatment for appendicitis is surgery to remove the appendix before it dangerously ruptures. When the appendix is gone, though, we are none the worse for it – so why do we have it in the first place? Other mammals have an appendix that is more developed than ours, the thinking being that it had a role in digesting plant-based foods. Just like we no longer need wisdom teeth, the function of the appendix is now defunct, and it remains as a vestigial remnant of our evolutionary development from primates. That said, evidence is building to support a role for the appendix in gut immunity, keeping a healthy microbial balance and fighting infection on the front line.

Other parts of the body seem to be evolving, too. The pinky toe is getting smaller and smaller; maybe it will soon be gone altogether. The so-called tailbone at the end of the spine, the coccyx, is the remnant of our evolutionary link with primates and their prehensile tails; you don't know you've got one until you fall on it! And that cute little protrusion on the rim of the ear helix is called Darwin's tubercle, a slight thickening that represents a much more prominent projection on a monkey's ear. These are just some of the physical features that make up our evolutionary archive.

Is tongue rolling a genetic trait?

Spoiler alert – no it is not! Tongue rolling, the ability to curl up the edges of your tongue to meet each other, is not any more likely to be seen in both identical twins than it is in fraternal twins. In fact, some non-rollers can even teach themselves to do it.

However, all is not lost. Some people have the even more extraordinary ability to roll their tongue around in several different configurations. Known as the cloverleaf tongue trait, this ability does seem to be dominantly inherited – if you can do it, at least one of your parents can probably do it, too.

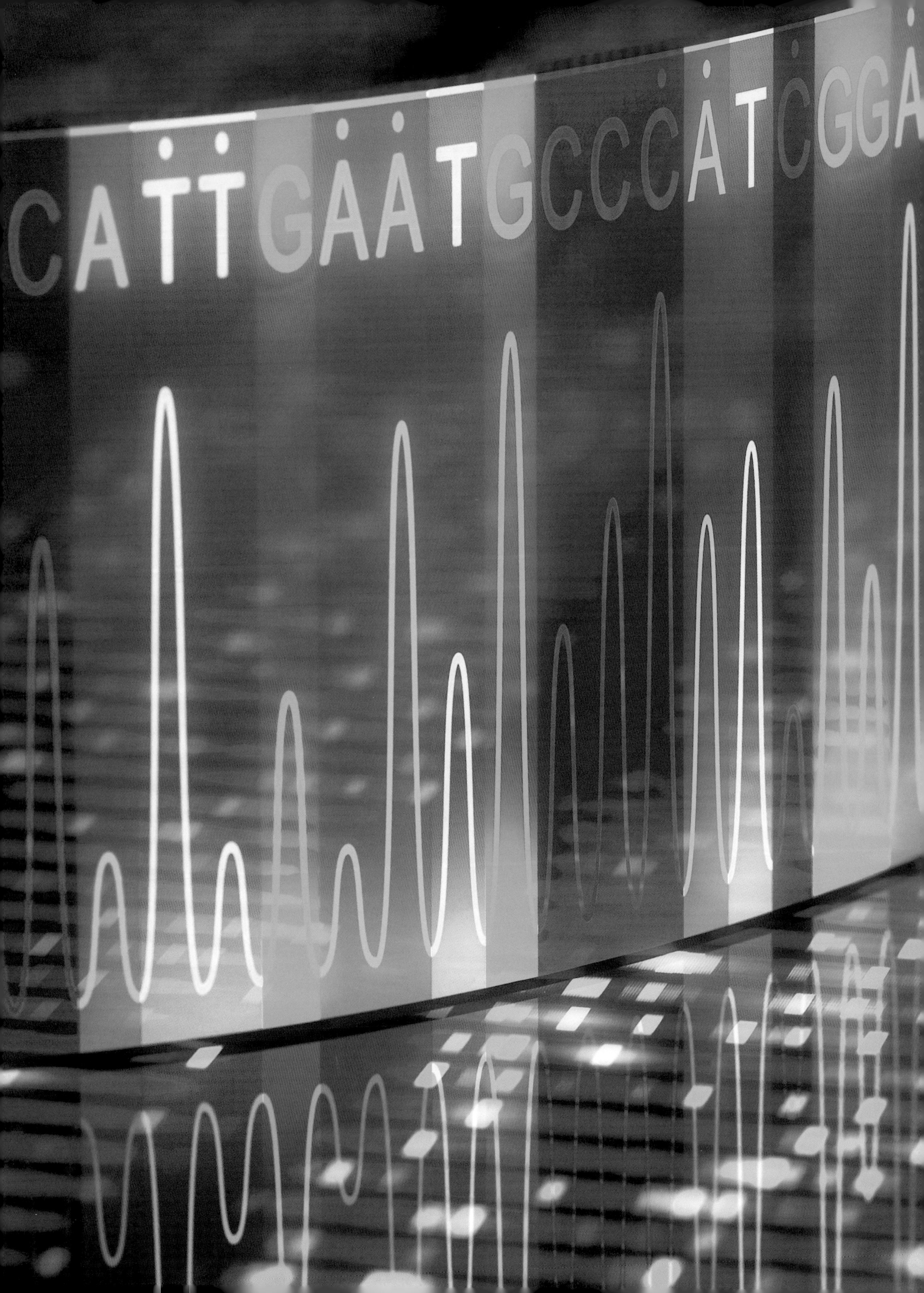
CATTGAATGCCCATCGGA

CHAPTER 5

How we use DNA today

There is a whole lot of information tied up in the blueprint of our genetic code. Unlock its secrets and who knows what you might discover about yourself (and what others can find out about you)? In this chapter we'll explore how we use DNA to benefit our lives and rock the world.

Direct to Consumer (DTC) genetic testing

The advent of a revolution in genetic testing technologies has paved the way for personal genome analysis. Known as Direct to Consumer testing (DTC), it is as simple as providing a saliva sample (cheek cells in your saliva provide the keys) and sending it away for sequencing. The results can explore several aspects of your lifestyle, health and ancestral background.

Health

Companies offering a personal genome service, like 23&Me (www.23andme.com), screen the DNA in your saliva for genetic variants associated with certain heritable conditions, aspects of health later in life and, if you want, your ancestry, too – all from the comfort of your own home. Testing will look for predisposition to a number of medical conditions that have some genetic link. This includes late-onset dementia; breast, ovarian and other cancers; macular degeneration that causes visual loss in adults; and a propensity towards blood clots – all by looking for certain genetic variants that scientific studies have shown to be associated with these and many other medical conditions.

When the results are ready, you are sent an email inviting you to log on to the website to access your report estimating the risk you have of developing one of the conditions. For some of these, such as inherited breast and ovarian cancer and late-onset Alzheimer's, the test looks for genetic variants with a stronger predictive power – if you have the variant, there is a high chance you will develop the condition, although it is not 100% likely. For other conditions in the panel of tests, the link between having the genetic marker and actually having a medical problem as a result is not as clear-cut – the risk estimates can be confusing, too. And while finding out that you have an increased risk of developing something means you can take precautionary measures for some conditions (such as Type-2 diabetes) that's not the case for every potential risk diagnosis.

Being able to take precautionary measures is one good reason for taking a test, but there are others. Carrier testing

▼ Genetic testing, previously the domain of the science laboratory, is now brought to you in your living room.

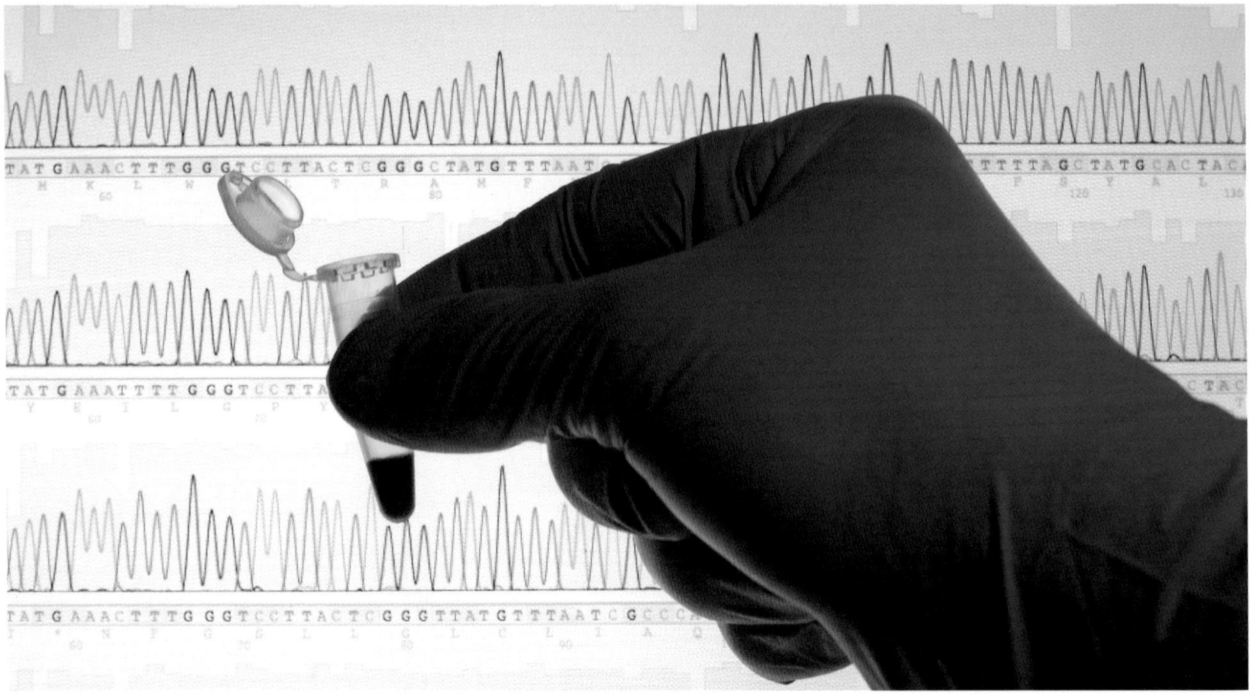

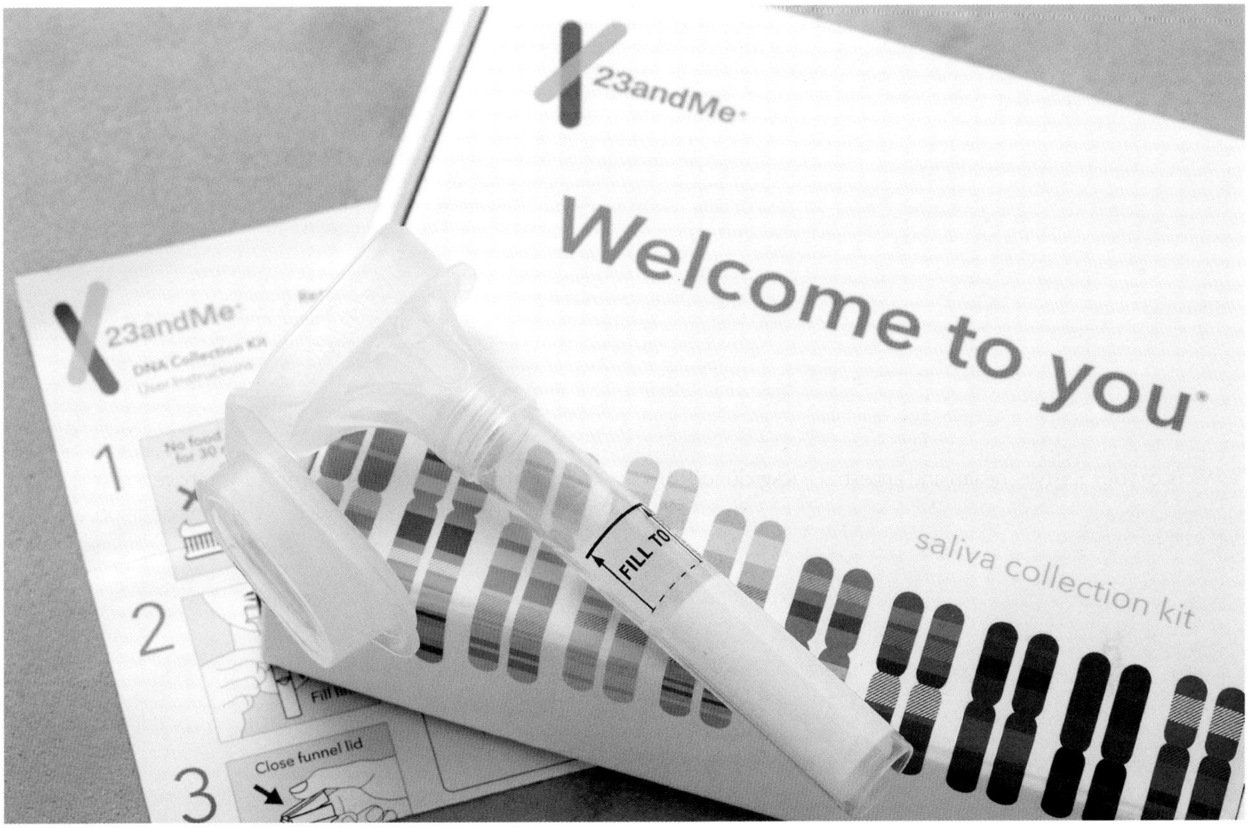

for recessive genetic disorders, such as cystic fibrosis and sickle cell disease, is useful if you want to think about what you might pass on to the next generation, if your partner is also a carrier. A woman can have screening to find out if she is a carrier of an X-linked genetic condition; and people from certain ethnic backgrounds can find out whether or not they are likely to carry conditions related to ethnicity.

Research

Twenty-six million people have used DTC DNA tests for ancestry and health. This equates to an unprecedented amount of genetic information, which, as a set of data, could help research efforts to find out more about the role genes can play in more common conditions, such as Type-2 diabetes. For example, scientists have discovered genes associated with being short-sighted (myopia) by asking people who submit their samples to provide information about their vision. New genetic links with the neurological movement disorder Parkinson's disease are also being researched in this way. Understanding the role genetics plays in causing debilitating conditions like this helps the development of new treatments, bringing us a step closer to personalised medicine, tailored to each individual.

For diseases like Type-2 diabetes (a disease that has reached endemic proportions worldwide), the DTC testing model creates research opportunities – scientists can compare the

▲ Companies like 23&Me offer direct-to-consumer genetic testing – all you need to do is spit in a pot!

▼ DTC testing can inform you about your health, your lifestyle and your ancestry by looking at your genetic code.

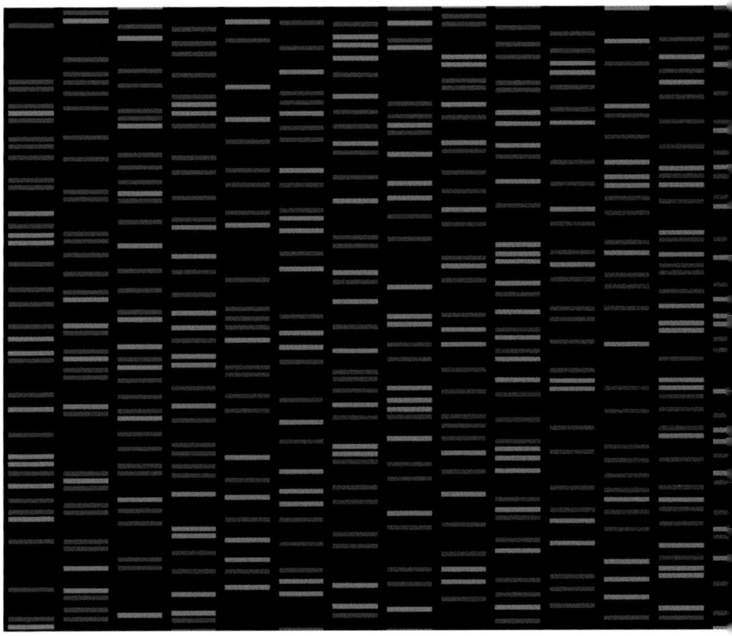

▲ **DTC genetic profiling offers to help you decide which lifestyle factors are beneficial to you.**

health data we submit with a susceptibility to the condition, as read through our ethnic heritage and genetic data. We can then explore the potential relationship between our DNA and environmental factors, such as poor diet and lack of exercise, in causing diabetes.

Ancestry

The concept of exploring ancestry has long been the domain of genealogists, who trace a person's lineage through finding information about their relatives and their pasts. Resources such as public records of birth, deaths and marriages, along with the national census, newspaper archives and immigration documentation, have long been the mainstay of their toolbox. Now, geneticists have jumped on the ancestry bandwagon, with DNA firmly placed in the repertoire of methodologies to trace your family's origins.

Ancestry DNA testing uses information in your genetic code to tell you what your ethnic origins are. This DTC test is based on markers in the genome found to be specific to a geographic region and therefore to the people living there (markers enable scientists to identify more than 500 different ethnic backgrounds). The test can tell you about the origins of your family's ethnic background several generations back – it can even look for Neanderthal DNA in your genetic code – but it cannot tell you about the more immediate past – where your grandmother was from, for example.

Surprises in your DNA

There is always a risk that digging through your DNA can turn up surprises that you may or may not want to have, or may reveal information that you'd rather keep to yourself. Siblings each taking a test, for example, could reveal non-paternity or the so-called NPE – 'not the parent expected'. Relatives of criminals, where the relatives have taken a test, can inadvertently help to solve crime: genetic profiles stored as a result of DTC have helped to solve a murder, providing possible matches with DNA found at the crime scene. Border control agencies have used stored genetic data to counter the claims of potential immigrants to right of entry by 'proving' genetically that the immigrants are not from where they claimed to be.

The test uses characteristic markers in your genetic code to build up a profile of where your ancestors originated. Different DTC tests use different methods based on autosomal, mitochondrial and Y-chromosome DNA. If you are curious to know which part of Europe your distant relatives came from or how many parts African you might be, the DTC test can help. And if you tick the 'connect me' box, you might even find living relatives you didn't even know you had.

Lifestyle

Under the banner of lifestyle genetic testing, you can have your genome analysed to inform you about what cosmetics suit you and which fitness and nutrition plans are beneficial.

DNA-based cosmetics

On the basis of a handful of markers revealing what lies under your skin, you can get beauty regimens and customised, skincare advice based on your DNA. This 'dermocosmetic' DNA technology uses your variant profile to build up a picture of your skin type. It aspires to inform you about elasticity, sensitivity, response to ageing, balance of oiliness to dryness, tendency to wrinkles and sensitivity to UV light.

DNA-based nutrition and fitness

Several companies offer DNA profiling that aim to reveal factors in your genome that might influence your optimal fitness programme, your body's response to dieting and your recovery times from sports injury. They use DTC testing to analyse a set of key markers in the genetic code, ones that have been shown through research studies to have an association with factors potentially relevant to a healthy and active lifestyle.

Nutritional genetic profiling can tell you, for example, how fast you metabolise carbohydrates and fats, and whether you have greater need for particular nutrients, such as omega-3, which can protect you from heart disease. Based on your DNA, you might need to reduce certain nutrients or stimulants, such as caffeine; and it could identify food intolerances, such as an intolerance to gluten.

You can also find out how your DNA holds the key to the way you should work on building peak performance and endurance, what (genetically speaking) your exercise capacity should be, how prone to injury you are and how best to recover after exercise. All of this gives you a bespoke training plan, based on your individual genetic profile.

▼ DNA-based cosmetics, where products are tailored to your skin, is a growing area in the beauty industry.

▼ Could your genetic profile point you in the right direction to a healthy and active way of life?

Should you have a DTC genetic test?

Genetic testing can be mind-bogglingly complex, with confusing results that you might not be expecting. In the medical setting, people who undergo genetic testing also receive genetic counselling – but if you buy a DTC, you're more likely to be left to your own devices. Try the following simple survey to see if you're really someone who should add-to-cart and spit.

Why do I want to take the test?

A **I'm laid-back – there's no particular reason.** Sounds like a bit of fun, that's all. Anyway it was a gift from my best mate, so it would be rude not to!

B **I'm quite interested.** It's so easy to get genetic testing done these days. Anyway, if there's something in my DNA I should know about, why wouldn't I want to find out?

C **I'm a real worrier.** And I am certain there could be something wrong. One day I'm going to develop a genetic disorder, I just know it. Worse, I could pass something on to my kids. How can I live without knowing what I'm facing?

What exactly is the test looking for?

A **Not really sure.** There's a list of conditions on the website, but no time to read the science bit. Anyway, the geeks in the lab have probably worked out what's important to know. Guess I'll find out when I get the results, eh?

B **I've read the small print on the website.** I've got an idea what they're testing me for – some late-onset conditions (some you can treat, some you can't), and carrier-testing too, which will come in handy before we have kids. Seems pretty reasonable.

C **I've Googled everything on the website list.** It's all really confusing, though – I've never heard of many of these things before. Although I have heard of Alzheimer's. It's good I am being tested for that because last week I couldn't remember where I left my keys and I think I've got it. Would be good to know either way – I can't face another sleepless night. Let's do this already!

OK, so you spat in the pot and sent it off – easy! But the results are due to arrive in your inbox any moment now. How do you feel?

A **Oh yeah, I'd forgotten about that.** Just let me finish watching this box set and I'll open the email.

B **OK, take a deep breath.** Could be good news, could be bad news. I'm just going to put the kettle on and prepare myself to open the report file.

C **Oh no! Help! It's here.** Wait, it's fine – I wanted to do this, remember? To find out if I've got Alzheimer's, right? Oh no, what if I've got Alzheimer's? I couldn't bear another sleepless night.

You've read the report and you're taking it all in. How are you feeling?

A **OK, so I'm 2.3 times more likely to get diabetes than the next guy.** Good to know. Right, which box set shall I watch next?

B **OK, so I'm 2.3 times more likely to get diabetes than the next guy.** What does that mean? Is that a high chance? Twice as likely sounds like it – is it, though? I need expert advice on that. Better find a specialist in genetic testing to ask for advice about these results.

C **Oh no, I'm 2.3 times more likely to get diabetes than the next guy.** That's why I'm so tired – I'VE GOT DIABETES! At least, I haven't got the other thing – what was that again? I can't remember.

Your friend asks you if you would recommend having the test. What do you say?

A **Yeah, why not?** It's easy to do and they've got a two-for-one offer on at the moment. You can even pay a bit more and see how much Neanderthal you've got in you (a lot, in his case!).

B **I guess so, it's good to know about the diabetes thing.** Although I've found out from the genetics clinic that it's not that high really. I just need to look after myself a bit better.

C **Yes! They should totally get it done.** Who wouldn't want to know that they're going to get diabetes? Really, who? I was lying awake thinking about it all night.

In the medical setting, people who undergo genetic testing also receive genetic counselling - but if you buy a DTC, you're more likely to be left to your own devices.

The verdict - how did you answer?

Mainly As.
You're laid-back. You take it all in your stride. You're the type who will take the test and deal with the result later. Box sets to watch!

Mainly Bs.
You're curious. You are informed, but you don't know how you'd react if anything showed up. You've got a plan, though, and you know where you'd go to get more information to help you understand the results.

Mainly Cs.
You're a worrier. You want to do the test to make you less worried, but chances are it could go the other way. Before you take the test, make sure you know exactly what it can and cannot tell you. And make sure you know where to go for additional support.

The small print

Here are some other things to think about when you take a DTC test:

- Has the test been approved by the FDA (the US Food and Drug Administration) or another official body for medical use?
- What is your personal information being used for?
- How is your information kept private?
- Who owns your DNA and DNA data?
- Can you request that your information is deleted at any time?
- Who else is your DNA data being shared with?
- Are you happy about these third parties having access to it?
- Can you be identified by them?
- Is your data encrypted?
- Will your DNA be used for anything else in the future that you do not know about?
- Might your DNA data be linked to information about you on social media, such as your profile?
- How long will your data be held?

Genetic curiosities

Some DTC genetic tests offer to reveal slightly more quirky characteristics about you, like whether your fear of heights is rooted in your DNA, or if you are genetically more prone to mosquito bites. Here are some other traits you can explore for yourself.

After a meal of asparagus, does your urine smell, well, asparagus-y?

Half of us say it does and half of us say it doesn't. So who is right? Your digestive system breaks down asparagus to make metabolites, one of which is called asparagusic acid. This is what gives asparagus pee its distinctive odour. But it is the ability to sense that smell that is a genetic trait. There is a variant in one of our genes that renders half of us anosmic – we cannot smell it in the urine. Why not suggest a show of hands the next time it is served at a dinner party to see the dominant smelling trait in full effect, though be prepared not to be invited back.

The photic sneeze reflex or ACHOO syndrome

Do you sneeze when you look into bright light or sunshine? Does your mother do the same? How about your kids? Then you may have autosomal dominant compelling helioophthalmic outburst (ACHOO) syndrome – also known as the photic (light) sneeze reflex. The Greek philosopher Aristotle (born 384 BCE), first identified people with this genetic trait when he asked in 'Problems concerning the nose', why is it that 'one sneezes more after one has looked at the sun?' Incredibly, we have yet to identify the genetic cause behind this peculiar phenomenon. Bless you!

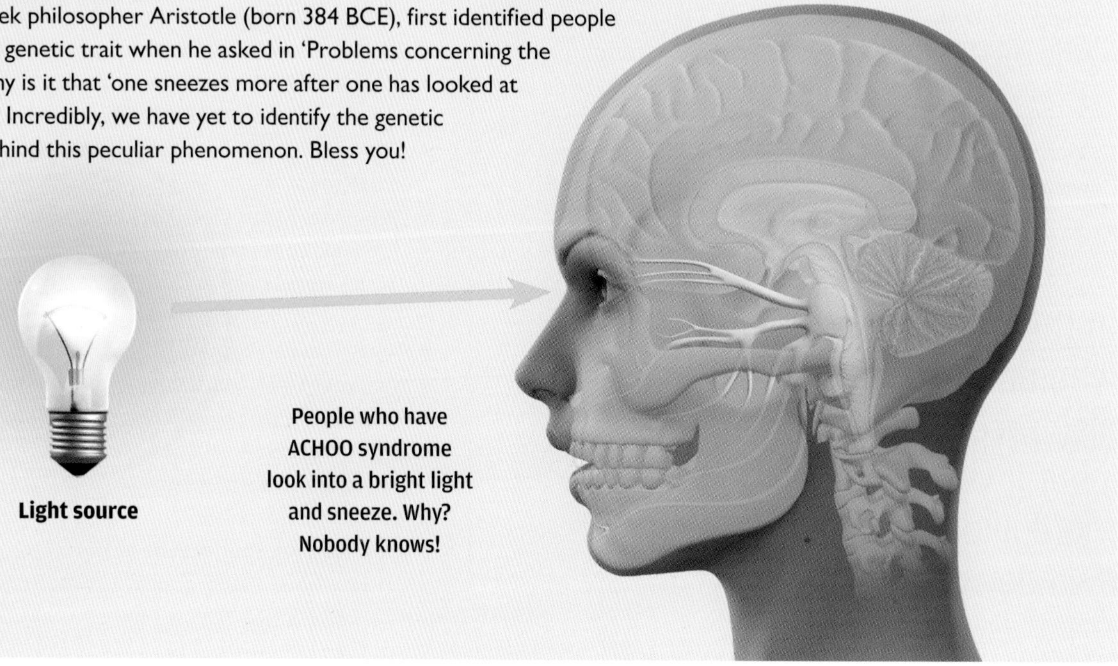

Take the cracker test

Do this: take one salt-free cracker, put it in your mouth, set a timer and start chewing. Do not swallow yet. Take note (in seconds) of when you start to sense a sweet taste.

What has this to do with your genes? Starch is a form of sugar found in carbohydrate foods such as bread, potatoes, flour and rice. When we eat starch, enzymes break it down into smaller sugars, and ultimately into glucose. One of the first enzymes to start this process is amylase, which is found in saliva. A chemical reaction, called hydrolysis, converts starch first into maltose, which amylase then breaks down into glucose, which tastes sweet.

Amylase is a protein and therefore made by a gene – in this case the gene *AMY1*. However, rather than having the usual two copies of that gene, one from your mother and one from your father, you can inherit multiple copies. If you have more than two, you might produce more amylase than other people. If this is the case for you, your body releases glucose from starchy foods quicker than someone with only two *AMY1* genes. The cracker test works out how long it takes your body to release glucose from a cracker – giving a sense of how many copies of *AMY1* genes you may have.

Analysing the results

- You tasted sweetness in under 14 seconds: you have the highest number of *AMY1* copies (maybe even more than 10).
- You tasted sweetness in over 30 seconds: you have the lowest copy number (maybe as few as only one).
- If your time fell in between: you have a number of copies in the middle range.

It is thought that numerical variation in our *AMY1* genes is a response of early humans to evolutionary pressures incorporating starch-containing foods into the diet, including tubers and root vegetables, just like the persistence of lactase production into adulthood (see pages 92–3).

Brussels sprouts – are you a lover or a hater?

Always getting told off for not eating your greens? Nightmare recollections of Brussels sprouts at Christmas and cabbage at school? Maybe you have the taste-receptor gene variant that means these and other cruciferous vegetables taste disgusting to you. They contain a chemical similar to PTC (phenylthiocarbamide), which was discovered when it was created as an artificial sweetener. For half the people who tasted it, it was horribly bitter because they have a different version of a taste-bud receptor on the tongue.

To test whether your hatred of your greens is genetic, you need some PTC taste strips, which are available online. Put a strip on your tongue and if it has a bitter taste, you are one of the 50% of people who have one of the three gene variants needed to detect PTC. Some of us even have a double dose of these bitter taste variants, inherited from both our mother and our father. If someone shows more exaggerated disgust to the strip, then they could well be one of these genetic *super*tasters. The same concept is thought to be behind why some people love Marmite and some hate it. In fact, the people at Marmite have begun the Marmite Gene Project, enlisting the help of a DTC-genetic test company to try to identify genetic markers that could explain the extremes of reaction in their consumers. The results are awaited with Marmite-y breath!

Genes and obesity

Increasingly, we live in a world that makes it easy for us to put on weight – we consume high-energy foods without burning off the excess – and in which the prevalence of medically defined obesity has tripled since the 1970s. Today, 670 million adults and 124 million 5–19-year-olds are classed as obese.

Obesity is defined as having a body mass index (BMI – the ratio of your weight to your height) of over 30, where the desired level is 18.5–24.9 and being overweight is 25–29.9. In evolutionary terms, our bodies were not so much designed to respond to times of feast, but to famine, so that nowadays people whose BMI is above 25 are more likely to develop a range of serious health problems, such as Type-2 diabetes, high blood pressure, heart disease, stroke, arthritis, obstructive sleep apnoea, cancer, kidney and liver problems, and even sudden death.

Lifestyle preferences have clearly played an important part in the rise of obesity to the endemic levels we see around the world today, but what role do the genes play and might our DNA hold the information we need to protect us from the health risks?

There are a handful of genetic syndromes, characterised by multiple symptoms, in which early onset, severe obesity occurs, suggesting that genes might well be involved. One such example is Prader-Willi syndrome, a disorder of gene imprinting where a child is missing the paternal copy of a critical part of chromosome 15 and instead has two copies from their mother (maternal uniparental disomy of chromosome 15). While children with this syndrome are difficult to feed as babies, they soon develop a voracious, insatiable appetite, even during the night, frequently eating some unusual foods, such as raw eggs, so powerful are the genetic drivers.

Other rare childhood syndromes in which extreme obesity occurs from very early on are now known to be associated with the disruption of two linked chemical pathways called the melanocortin-4 receptor (MC4R) pathway and the leptin pathway. Research has identified a number of genetic variants that change proteins in these pathways and disrupt regulation of a healthy weight, a setting that was chosen in response to evolutionary pressures. In this respect, the important genes in those pathways include *LEP* (leptin), *MC4R* (melanocortin-4 receptor) and *POMC* (pro-opiomelanocortin).

The BMI index

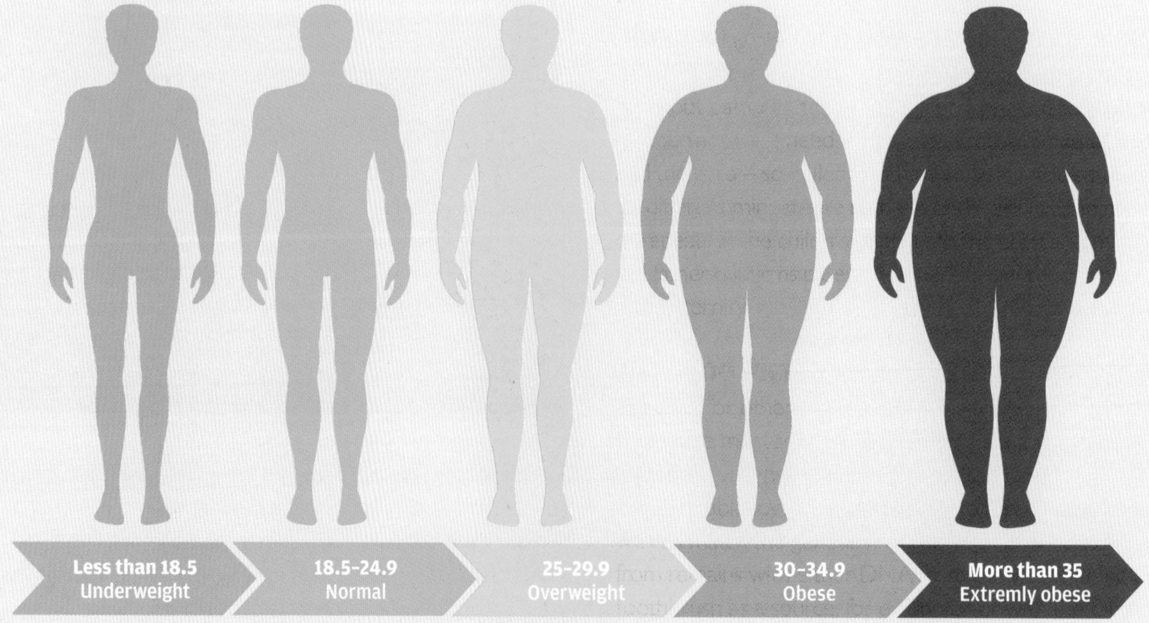

Leptin

Leptin is a hormone produced by adipocytes, cells that store excess fats and lipids that are found in the fat layer under the skin, called the adipose tissue. It inhibits the satiety centre in the brain's hypothalamus, so that we don't ever feel full. Leptin levels reflect the amount of fat stored in the adipose tissue – the more stored fat there is, the more leptin the body produces. This is thought to be an effect of evolutionary forces – the body uses the fat as fuel to protect us against starvation when food sources are scarce.

Children with the autosomal recessive condition congenital leptin deficiency have severe obesity – the body thinks there is no leptin, not because of a genetic defect preventing it being made, but because it mistakenly believes there is no fat. So the satiety centre is constantly switched off, resulting in over-eating and excessive weight gain.

Melanocortin-4 receptor (MC4R)

The MC4R pathway is responsible for regulating body weight. The receptor is found in specialised nerve cells in the brain. When it is activated, it fires off signals to control both energy intake and energy expenditure to reach the right balance. Dominant and recessive changes in the *MC4R* gene can cause MC4R-related obesity. In fact, such changes make up the biggest single gene cause of obesity, being responsible in 5% (1 in 20) cases of people with severe obesity. Other effects on the body of MC4R-related obesity include increased lean mass – your body mass minus the fat mass, which is derived from your bones, muscles and internal organs – and increased growth rate in childhood. This can create the effect of being 'big-boned'.

Pro-opiomelanocortin (POMC)

A protein produced by the pituitary gland, POMC is made up of different proteins joined together. When the full-length POMC is broken up, it produces three different, shorter proteins: adrenocorticotropic hormone (ACTH), endorphins and melanocyte-stimulating hormone (MSH). ACTH, a hormone produced by the adrenal gland in the abdomen, is responsible for catabolic steroid production in the body, while endorphins activate the body's opiate receptors, producing a natural pain-killing effect.

MSH exerts its effect by triggering the MC4R receptor. Disruption or deficiency of the MSH portion of POMC results in POMC-related obesity, because we never feel full.

FTO

The fat mass and obesity-associated gene, *FTO*, was the first to be identified as containing genetic variants associated with increased BMI and a tendency towards becoming obese later in life. There are five known variants associated with the *FTO* gene, and they can all be found in close proximity, which means they are often inherited as a 'block' of markers – so if you have one it is likely that you will have them all. One particular variant, the rs9939609 marker in which a T is replaced with an A, is associated with, but not necessarily causative of, an increased chance of being obese. People with two A-alleles (AA) have a higher chance of this than people with AT.

It is still not clear why variants in *FTO* should result in a tendency for obesity – it might be mediated through the brain and binge-eating behaviour, or it could be related to how the body metabolises carbohydrates. The AA form of the gene is associated with a 2.5 times higher chance of obesity when eating a high-carb diet compared with people with the TT allele. Knowing that you have these variants does not mean you will definitely become overweight, but it allows you to take steps to modify your lifestyle to stay healthy, and some of these markers are looked for in DTC health and fitness genetic tests.

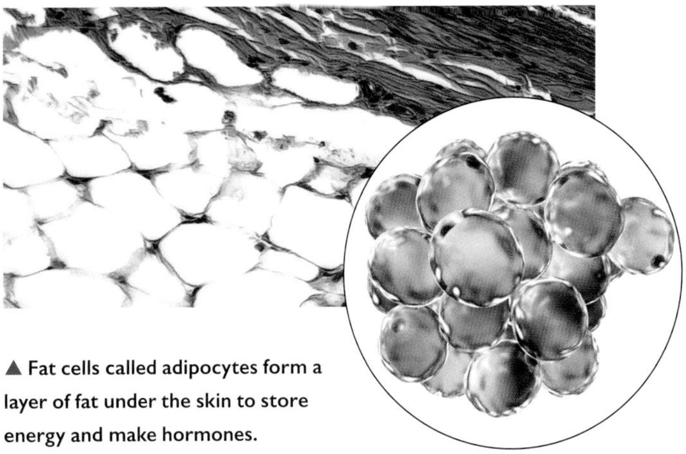

▲ Fat cells called adipocytes form a layer of fat under the skin to store energy and make hormones.

Polygenic obesity score

Studies show that – overall – several genetic factors influence becoming obese and can explain 40–70% of the difference in propensity to it from one person to another. Leptin levels do seem to play a role in 'everyday' obesity, some people having abnormally high levels. Those people may even develop leptin resistance, where the body no longer responds to the hormone (a bit like insulin resistance in diabetes), so that the brain doesn't register that the person is full.

The cumulative effect of having a number of the genetic obesity factors, each exerting a small effect, means doctors can give an individual a polygenic obesity score – a calculation that predicts the risk of developing a gene-related obesity condition.

The genes of a long life

Queen Victoria was 82 years old when she died in 1901 – not a bad effort during an era when the average life expectancy for women was 48. Queen Elizabeth the Queen Mother was 101 when she died. When Akihito, the 125th Emperor of Japan, abdicated in April 2019, he was aged 85. He had succeeded his father, Hirohito, to the Chrysanthemum Throne on Hirohito's death at the age of 87. His son, Narohito, must have been preparing for the long haul as he acceded to his current position of Emperor in 2019, at the age of 59. When you look at families like these you cannot help but think that living to be an old age must be genetic.

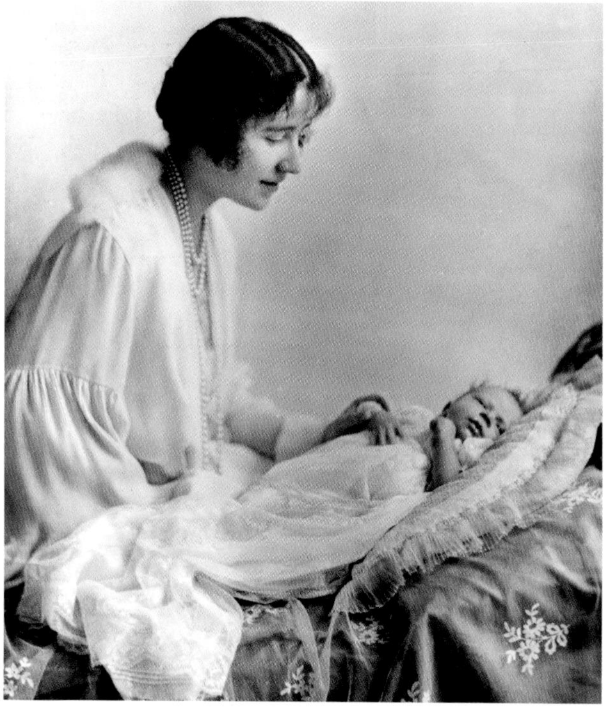

▲ **Queen Elizabeth, the Queen Mother, in 1926 with her daughter, the future Elizabeth II.**

Longevity genes

Researchers looking for genes that might account for longevity studied people aged over 110 years old, known as supercentenarians. They discovered a number of genes in these remarkably elderly people that seem to be markers associated with exceptional longevity, like *FOXO3* (forkhead box O3 – a transcription factor important in renewing cells). The genes they found could be placed in two distinct groups: cell maintenance and protection against certain illnesses.

Cell maintenance

Genes responsible for the maintenance of cells can perform such functions as repairing damaged DNA, maintaining the length of telomeres (the ends of the chromosomes that get shorter with age), and protecting against the deleterious effects of free radicals ('rogue' cells that cause ageing and disease). Genetic studies suggest that supercentenarians have genes that carry out these tasks far more effectively.

◀ **Japanese Emperor Hirohito (centre) and the Imperial Family, including his wife, Empress Nagako (seated at left) and his son, Akihito. A family in which incidents of longevity may be due to their genetics.**

Disease protection

The genes in the second group have markers that are linked with an increased risk of certain health problems, such as heart attacks, stroke and Type-2 diabetes, at least when they occur in people of average life expectancy. When an alternative set of markers is present, they seem to be protective instead – we can't go as far as to say they definitely are protective, just that in those without the markers of increased health risk, no news is good news.

The consequences of this are apparent only in the exceptionally old because, before then, environmental factors (such as smoking and an unhealthy diet) seem to have a more pronounced effect upon health (and therefore survival age) than genetics.

Heritability of longevity

However, DNA is not the only explanation for longevity. Another team of researchers surveyed 540 million family trees that were submitted to the Ancestry DNA consumer-testing service. The researchers concentrated on individuals who were born in the 1800s and the early 1900s, but who are now deceased, focusing on their age at the time of death. When this information was compared with information from first-degree relatives, such as parents and siblings, genetic factors appeared to explain 20–30% of differences in longevity, in keeping with previous studies showing 25% heritability.

The Ancestry DNA data also showed that spouses tend to die at a similar age – which is not surprising perhaps given that couples share non-genetic influences on their longevity because they live together. What was surprising, though, was the observation that non-blood relatives, like siblings-in-law and first cousins-in-law, also died at similar ages – an observation that took the heritability of lifespan down to only 7%. The reason might be assortative mating, the concept that people chose partners who are similar in nature to them.

Lifespan score

Scientists at the University of Edinburgh studied the DNA of 500,000 people and asked them to what age their parents had lived. Combining the information allowed the researchers to identify 12 regions in the genetic code that seem to influence lifespan. They came up with a lifespan score based on these polygenic links. When they ordered the participants into 10 different groups according to their scores, the scientists discovered that the top group were predicted to live as much as five years longer than the lowest-scoring group.

In truth, then, it seems that how long we live is a genuine combination of both genetic and environmental factors. So, if you want to max out your chances of receiving a Royal telegram on your 100th birthday, don't smoke, do eat and exercise well – and choose your partner carefully!

The family tree

Constructing a family tree, also known as a pedigree, shows us where we fit in amongst our relatives and ancestors. It is a useful diagram for geneticists and genealogists alike, helping track familial traits and finding your place in the family. It could even reveal a surprising past. Who do you think you are?

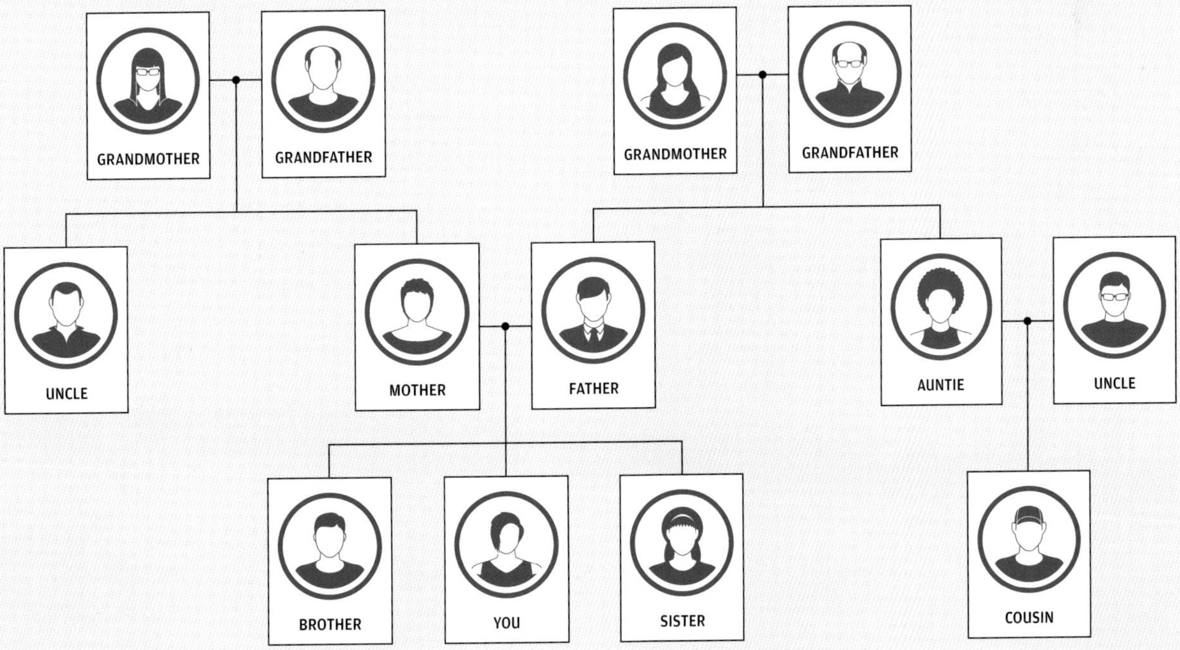

The genes of intelligence

Studies involving twins – both identical and non-identical – show that intelligence is a highly heritable trait. In fact, depending on which study you look at, intelligence is 50–80% heritable, meaning that genetic factors account for most of the differences in intelligence from one person to the next. Intelligence is multi-faceted – there is no single defining characteristic and many distinct qualities could make us call someone intelligent. With this in mind, how can we begin to find out if what we call intelligence is genetically determined and what lies beneath it?

One relatively objective measure of intelligence is called your IQ (intelligence quotient). An IQ assessment uses the Wechsler Adult Intelligence Scale (WAIS) to calculate a score. A full-scale WAIS test is made up of different subtests that draw out different aspects of intelligence, such as verbal intelligence and performance (perception and processing skills). Verbal IQ, which measures acquired knowledge and the ability to reason and problem solve, seems to be much more highly correlated with underlying genetic factors. Assortative mating potentially features again here, by adding to the heritability 'load' – like-minded people being more likely to choose one another.

The trouble with science

Fundamentally, it's very difficult to conduct genetic studies relating to the heritability of intelligence. Some studies have used the GWAS technique to analyse people across the genetic code who have high educational attainment. That is, the studies have looked at those who have spent the greatest amount of time in a learning environment, which seems to work best in scientific studies as a 'trait' to link with genetic background. Certainly, a whole host of genetic markers of intelligence have been found this way, but the results reveal major limitations. First, one study does not always replicate the findings of another and reproducibility in science is key to establishing facts. Second, we would need a huge cohort of people, numbering millions, to generate a list of genetic variants associated with intelligence when the effect of individual markers is so tiny. To find something of true relevance requires massively up-scaling these studies – which is often logistically prohibitive.

That isn't to say, however, that existing studies aren't valuable – there are studies that have been large enough to have statistical power, which did, to an extent, give reproducible results and identify new genetic markers associated with intelligence. In fact, there are over a thousand such studies and making sense of them all will prove difficult, especially as only a small proportion of markers are found in regions of the genetic code that actually produce proteins.

These markers, though, are expressed in parts of the brain that are known to have links with decision making, executive functioning, problem solving and memory. In which case, researchers seem to be on the right track, it's just that there is a lot more to do to get to the bottom of it all. And, until we do, we have to assume that, like so many other human traits, intelligence is primarily the product of both genetic and environmental factors. In other words, educational opportunities, parenting styles, nutrition and socioeconomics – among myriad other things – play an important part in the development of cognitive ability, too.

◀ **IQ (intelligent quotient) tests measure different areas of ability.**

The genes of gender

Male and female sex determination is under genetic control. Males carry sex-determining genes on the Y-chromosome, such as the *SRY* gene. Foetal production of the androgen hormone testosterone drives differentiation into a male, as does anti-Müllerian hormone, the presence of which results in the reversal of the organs that would otherwise become the female uterus and tubes. How this relates to gender – the binary social and cultural differences between the sexes rather than biological ones – is not obviously genetic. Copy number changes of the sex chromosomes do not seem to have an association with gender. Men with an extra sex chromosome, like XYY males, are not more masculine than their XY counterparts; and XXY males are not more feminine for having an extra X chromosome. But how about sexuality? Could sexual preferences be genetic?

Genetics and sexuality

Genome-wide association studies (GWAS) of roughly a thousand men self-identifying as homosexual in comparison with a thousand men self-identifying as heterosexual flagged up two markers that seemed, in this relatively small number of subjects, to be related to male sexual orientation. The markers were detected in *TSHR*, the gene that encodes the thyroid-stimulating hormone (TSH) receptor, and *SLITRK6*, a gene rarely linked to congenital deafness. Why markers possibly linked with sexuality were found in genes seemingly unrelated to sexual preference is not at all clear. And genetic studies involving twins have thrown up results that potentially conflict, some being suggestive of an underlying genetic link, and others, centred around identical twins discordant for their sexual preferences, conclude that a genetic link is unlikely.

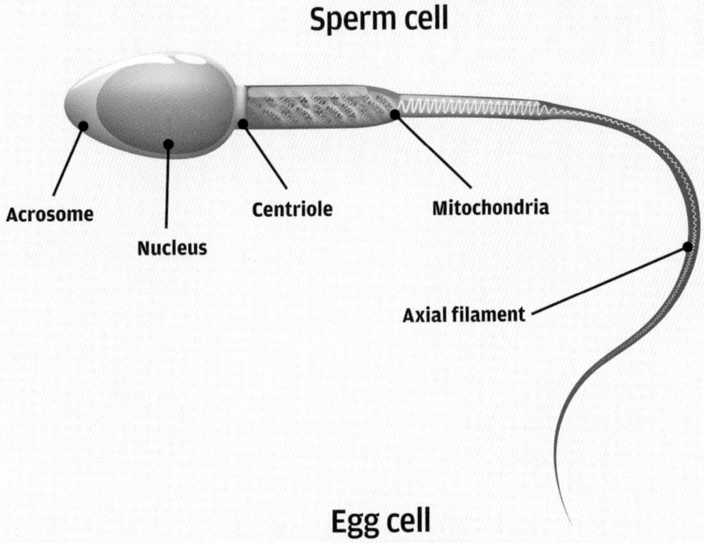

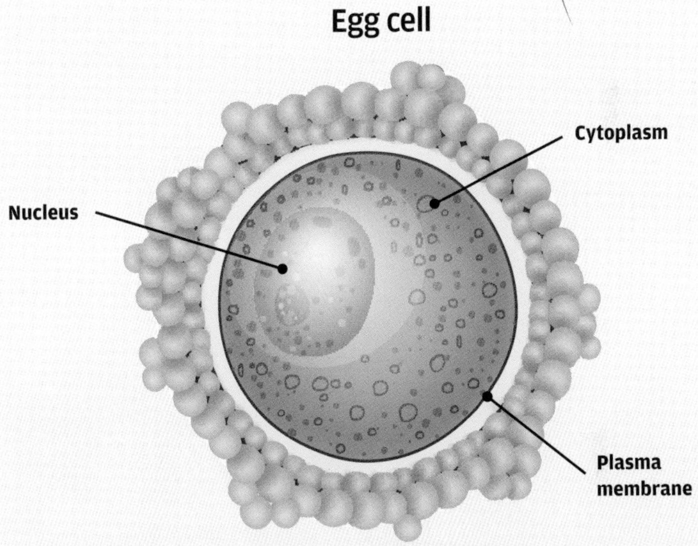

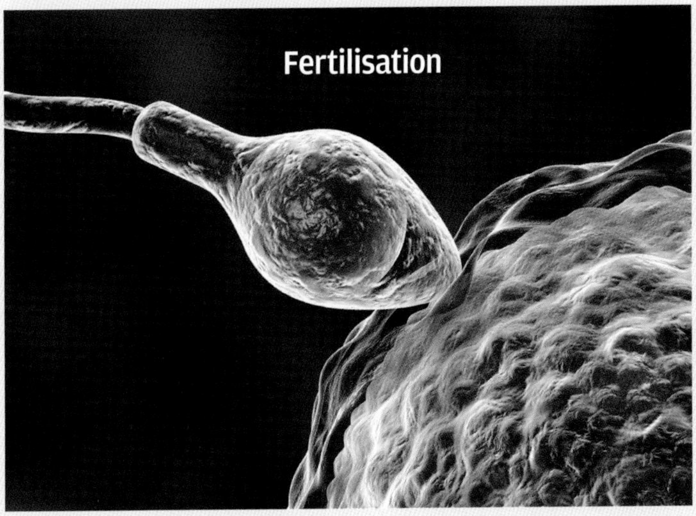

The genes of strength and stamina

One look at marathon medals' table shows that the sport is dominated by Kenyan runners in both the men's and women's competitions. Of the top 10 fastest male runners of all time, five are Kenyan. And all the others come from the neighbouring nation of Ethiopia. The same is seen in the women's table: six of the fastest originating from Kenya and three from Ethiopia (the UK's Paula Radcliffe is top). Most of these runners come from one particular tribe of 5 million people called the Kalenijn. Likewise, the small island of Jamaica, with a population just shy of 3 million people, contributes a disproportionately high number of Olympic medalists to the sprint disciplines of 100m and 200m. Take Usain Bolt, Yohan Blake and Shelley-Ann Fraser-Pryce for starters. So, is there something genetically special about the athletes from these groups that makes them so successful in their fields?

Genetic effects on muscle fibres

Physical traits that might lie behind elite sporting success include the nature of the individual fibres in muscle and the body's ability to deliver oxygen to the tissues during strenuous exercise. Skeletal muscle, the kind that contracts to move joints, is made up of two different types of fibre called slow-twitch and fast-twitch. Slow-twitch fibres contract slowly and are most useful during prolonged physical activity, whereas fast-twitch fibres contract quickly, which is helpful for sprinting and for sports requiring strong muscles.

Two specific variants in the *ACTN3* gene, which makes a fast-twitch protein called alpha-actinin-3, have been linked

▼ The Jamaican men's 4X100m relay team, champions at the Rio Olympics, 2016.

▶ **Scientists have shown that high altitude mountaineers are assisted by their genetics.**

with different physical abilities. One variant, R577X, results in a shorter, truncated protein because the variant terminates transcription too soon, after the amino acid at position 577 is incorporated. Some people have this variant on both copies of *ACTN3* – known as 577XX – which means they have many more slow-twitch fibres compared to fast, making them really good at endurance events such as long-distance running. The opposite of this is having 577RR, with many more fast-twitch fibres. In this case, athletes have exceptional speed and strength making them especially good a sprint events, such as a quick 100m dash.

Another gene influencing athletic performance is *ACE*. This encodes the angiotensin-converting enzyme (*ACE*) responsible for changing a hormone called angiotensin I into angiotensin II, which stabilises blood pressure. (You may have heard of *ACE* inhibitors, which can be prescribed to treat high blood pressure.) However, it does more than that – markers in *ACE* have been linked with muscle strength and physical endurance as well. One marker, called the D-allele, which indicates a deletion of 237 nucleotide bases in *ACE*, raises *ACE* levels so that those with two D markers (the DD pattern) have the highest levels. This chemical trait is associated somehow with a greater number of fast-twitch fibres and, therefore, sprinting ability. The other marker, the I allele, causes a reduction in *ACE* levels and therefore lower enzyme activity. A team of scientists from the Xtreme Everest Research Group found that mountaineers who can climb higher than 8,000m are far more likely to have two I markers (the II pattern) with lots of slow-twitch fibres and improved physical endurance.

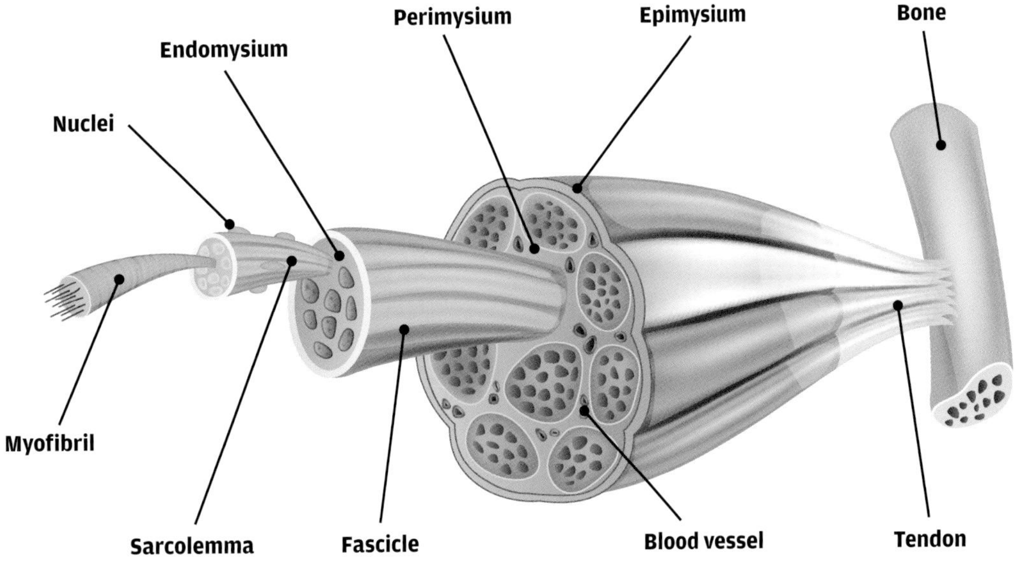

▶ **Skeletal muscles attach to bones, making us move when they contract. Each one is made up of many individual fibres. Genetic factors in elite athletes have been shown to improve their efficiency.**

Heritability of sporting success

A number of twin studies from as far back as the early 1970s show that certain physical traits compatible with sporting performance have strong heritability, meaning that there are strong genetic factors influencing success. For example, one study measured the VO2 max (physical endurance during aerobic activity) of identical and non-identical twins. The study showed that in 93% of cases heritability was a significant influence in physical capacity during exercise.

Tennis superstar sisters Serena and Venus Williams; English footballing legends, brothers Bobby and Jack Charlton; and the father-and-son goalkeeping duo, Peter and Kasper Schmeichel all indicate that sporting success does have at least some element of keeping it in the family. But does DNA have all the answers (and all the reasons) or is there some nurture involved? Perhaps their environments, including the family 'habit' and belief in training competitors to the top of their game, played more of a role – is it just that these athletes have grown up in an environment that supported the notion of practice, practice, practice?

Practice versus potential

Famously, Swedish psychologist K. Anders Ericsson concluded from his research that 10,000 hours of practice over a 10-year period was what it takes to be expert at something. It seems likely, then, that elite sporting performance is in general the product not of strongly influential single-gene effects, like *ACTN3*, but of multiple genetic factors that produce the potential to excel (nature) – *and* that nature's potential is realised only in the face of hard graft and performance development (nurture). Furthermore, Janet Starkes, a scientific researcher in body movement working in Canada, has demonstrated that sporting prowess relies on using specific cognitive skills, too. Brain as well as brawn, if you like.

▼ Athletes, including several sets of siblings, parade during the opening ceremony of the London 2012 Olympics.

▼ Tennis superstars, and sisters, Venus and Serena Williams.

Sex and sport

In 2014, the International Associations of Athletics Federation (IAAF) banned 18-year-old Indian sprinter Dutee Chand from competition because her testosterone levels (her levels of the 'male' or androgenising hormone) exceeded the organisation's guidelines for a female athlete. The IAAF explained that women with levels of testosterone sitting at the lower end of normal for men (identified as 10 nanomols per litre) have an unfair advantage as a result of increased muscle mass and strength.

The guidelines had been introduced to detect doping, when athletes take artificial hormones, such as testosterone, to improve their performance. What they had not reckoned on was finding female athletes, like Dutee Chand, whose bodies naturally produced increased levels of testosterone.

Dutee has the male chromosome pattern of 46,XY. Her body produces testosterone (all women's bodies do to an extent), but a genetic alteration means that it doesn't detect the testosterone, which in turn doesn't then have an androgenising effect on her physical characteristics – that is, Dutee doesn't have male genitalia (she has a female body with male hormone levels). This has been termed a 'difference in sex development' (DSD). Insensitivity to androgen hormones can be complete or partial and its physical effects can be different from one person to the next.

In 2015, Dutee challenged the IAAF's stance on testosterone levels in women. The Court of Arbitration in Sport (CAS) ruled that there was not enough evidence behind the claim of a competitive advantage to make such a blanket regulation. Evidence from subsequent research suggested that higher testosterone levels can improve performance by up to 3%. Two years later, then, the IAAF produced another guideline, this time specific to athletes competing in women's competitions with 46,XY DSD. Whereas the original guidelines did not allow women to compete if they had testosterone levels of 10 nanomols per litre or higher, the revised ruling reduced that level to 5 nanomols per litre, more in keeping with the normal female range. Hormonal manipulation using some types of oral contraceptive pills can regulate the levels, bringing them down into the Federation's acceptable range.

Another high-profile track athlete, South Africa's Caster Semenya, also challenged the IAAF, this time on their new 5 nanomol per litre ruling. And after an initial rebuttal, her challenge was accepted and female athletes with DSD will not be required to take medication for the purpose of competing.

▼ Dutee Chand during a training session four years after she was forced out of the Asian Games in a row over her gender.

▼ Caster Semenya competes in the Women's 800m race during the IAAF Diamond League event on 3 May 2019.

DNA profiling

The scientific study of the lines, arches, loops and whorls on our palms and fingertips is called dermatoglyphics (literally, what is 'carved' into the skin). In the early days of medical genetics, doctors and scientists pored over these lines, believing that their patterns could be linked with a genetic abnormality. Not that far removed from the imprecision of palmistry – the practice of characterising a person or foretelling their future by looking at their hands – reading the lines on our hands in a scientific manner has not been that useful in clinical practice. It is, though, much more effective for catching criminals and imposing border control.

Differences in our genomes are as individual to us as our fingerprints – and that forms the basis of DNA fingerprinting and profiling.

▼ DNA found in biological samples left at a crime scene can be used to catch criminals.

Catching criminals

In the interests of solving the crime and avoiding miscarriages of justice, DNA technology has been a gift to forensic science. The discovery that certain patterns within our genetic code were unique, just like fingerprints, meant that biological samples left behind at the scene of a crime provided a key that could unlock the perpetrator's identity. But, how?

Minisatellites

Minisatellites are sections of the genetic code where the DNA sequence is repeated over and over again, up to 50 times. There are over a thousand of these regions, each containing 10 to 60 nucleotides (those A,T,C and G bases). When minisatellite DNA is copied from one generation to the next, the number of times the sections are repeated changes. So, each new person has a unique number of repeats – with the exception of identical twins. (Out of interest, even identical

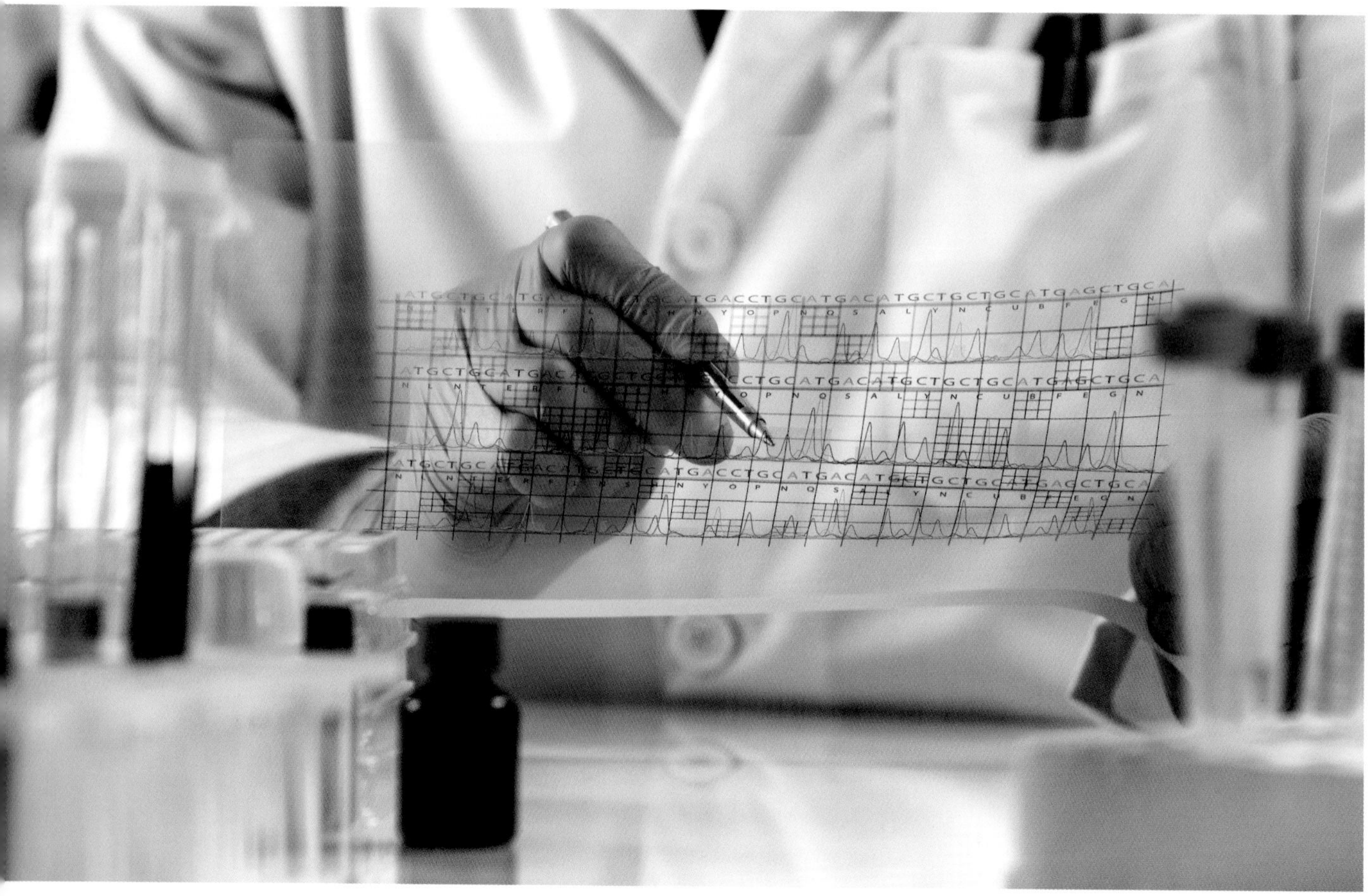

twins do not have the same actual fingerprints.) Using the RFLP method, forensics teams can use special enzymes to 'cut' out the DNA either side of the stretches of variable repeats, and then measure the size of the fragment to create a unique bar code of all the minisatellites.

Microsatellites

Nowadays, though, forensics teams can study repeated DNA segments much shorter than minisatellites. Microsatellites are short repeats in tandem – also known as short tandem repeats or STRs – two to five bases long. About 20 of these tracts make up the number needed for efficient DNA profiling. Creating a barcode of microsatellites is more efficient when dealing with the tiny samples of DNA that might be left at a crime scene. It uses polymerase chain reaction (PCR) technology too, rather than the clunkier RFLP enzyme method used to generate the relatively larger minisatellite fragments.

National DNA databases

Forensic scientists in the UK use the the DNA-17 system of genetic fingerprinting to make a barcode of 16 microsatellites and one that determines sex (DNA from the X or Y found in

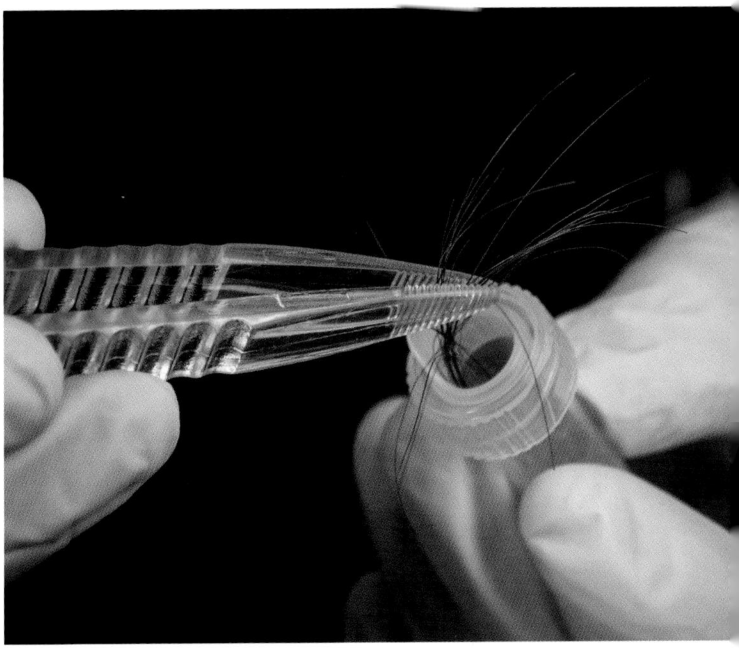

▲ **DNA can even be found in the smallest samples of hair and used for forensic profiling.**

DNA profiling

DNA profiling uses a similar method to Sanger sequencing. It takes a tiny amount of DNA, increases its concentration several times over (amplifies it) and then analyses the microsatellite markers at 17 different places, or loci. This builds up a DNA profile that is specific to an individual, which can be used to compare with other profiles to find a match.

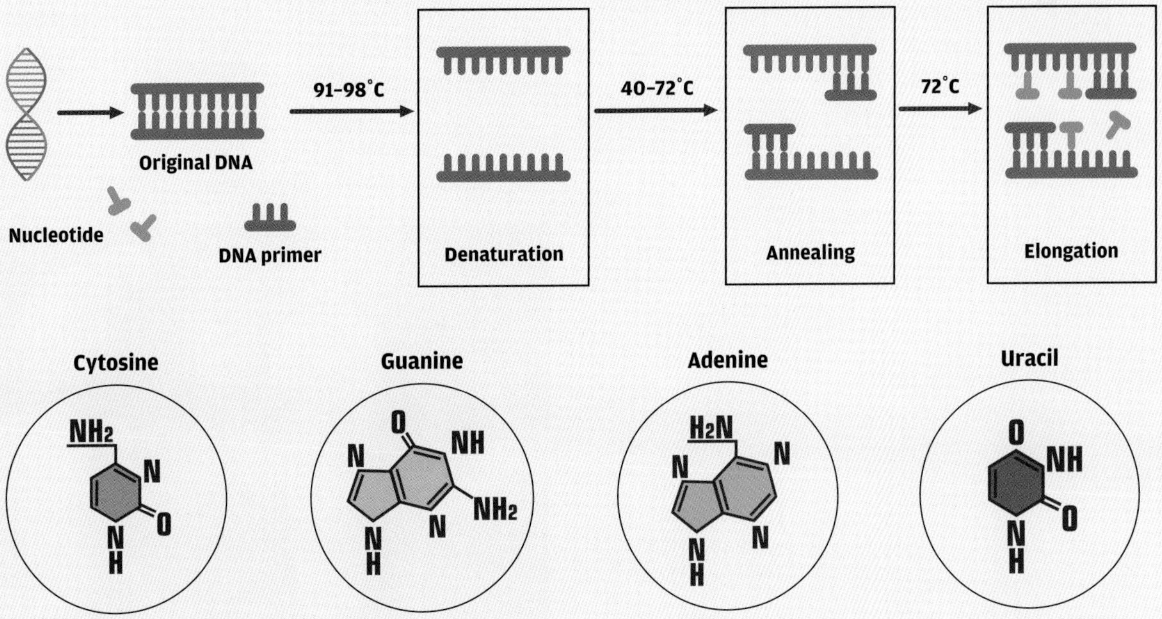

Alec Jeffreys (1950-)

In 1984 , Alec Jeffreys, a geneticist at the University of Leicester, realised he had stumbled across what turned out to be DNA fingerprinting while he was studying normal variation in human genes. He noticed that there were a different number of repeats from person to person, even if they were closely related. He concluded that even though we inherit one set of markers from our mother and one set from our father, the variation in the number of repeats was unique to a new human being in the same way that fingerprints are.

Crossover and recombination happens before the markers are passed on in the egg or sperm. While the combination might be different and unique, the individual mixture of markers can be identified as having come from one or other parent. Armed with this knowledge, shortly after his discovery, in 1985, Jeffreys helped resolve an immigration dispute by using genetic fingerprinting to prove that a young boy was genuinely the son of the people claiming to be his parents. Then came the first paternity test using DNA fingerprinting; its use in forensics was developed soon afterwards.

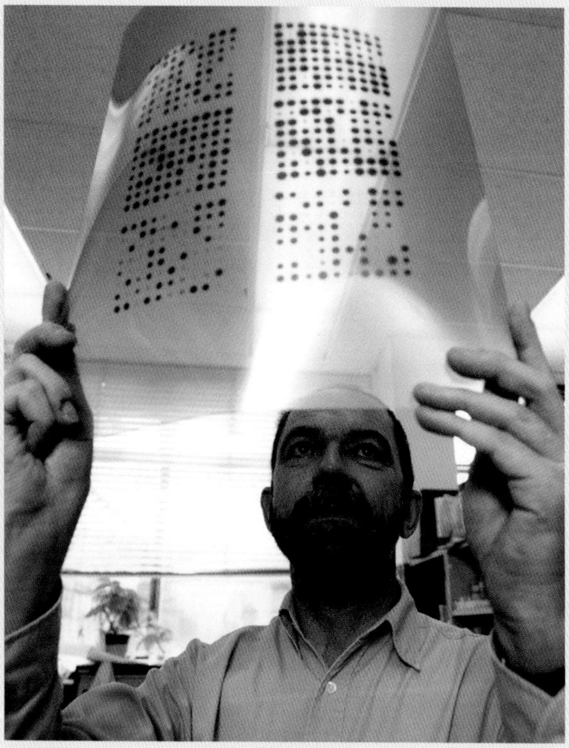

the *Amelogenin* gene). This generates 32 read-outs for each person, one from each allele and one that determines sex – information that's kept to help solve crimes. Most countries have a national database of DNA profiles. The UK database was established in 1995. Now under the jurisdiction of the Home Office, it keeps a permanent record of the DNA profiles of around 6 million UK individuals linked with serious crimes. Following a ruling by the 2013 Protection of Freedom Act, nearly 2 million profiles were removed.

DNA databases mean that as soon as new DNA evidence comes to light there is the potential for solving not only live cases, but also cold cases (ones unsolved for many years). While in the main that might mean catching criminals who have until now walked free, it also means that we are able to release or exonerate those who are wrongfully accused.

Crime scene investigation

The value of using DNA in forensic science is unquestionable – but just how useful it is comes down to accuracy, especially the care with which it is collected. Whole cases can rest on the precision of forensic examiners as they gather up samples from a crime scene; any hint of contamination can render DNA evidence inadmissable.

It's thought that DNA can remain on clothing after repeated laundering and that DNA from different people can even cross-contaminate inside the washing machine. There are even questions about whether everyone sheds on average the same amount of DNA from, say, their skin; or if some people, the so-called 'higher shedders', do so more than others. How much contact is required for an adequate amount of DNA to be left behind is also largely unknown. So, while the role of DNA in forensic science is well established, the science behind collecting it is still really in its infancy.

▼ Police forensic officers inspect the aftermath of a suspected car bomb explosion in Derry, 2019. (Getty)

The genes of criminality

DNA can help to solve crimes, but can it create the criminals in the first place? Several genetic studies would suggest so – for those with extremely violent behaviour, at least. Scientists seeking to identify potential genetic drivers of criminality have found related variants in a small number of genes.

The 'warrior gene' et al

A meta-analysis that combined all the results of 38 studies published between 1996 and 2006 linking antisocial, violent or aggressive behaviour with genetic influences showed that genes can account for 56% of the causative reasons people turn to crime (with 42% relating to environmental factors, such as growing up with violence in the family). Other studies have found specific examples of genetic variants that seem to link DNA and crime, especially in people committing extremely violent crimes. For example, a GWAS study in Finland found two genetic markers in the genes *MAOA* and *CHD13* in criminals who committed more than 10 counts of murder, attempted murder and battery.

MAOA makes a protein called monoamine oxidase A, an enzyme that metabolises dopamine. Dopamine is a neurotransmitter, a chemical messenger in the brain. It is released in response to a positive stimulus, a kind of chemical reward to reinforce 'good' behaviour. The *MAOA* variant found more commonly in violent offenders results in slower dopamine metabolism, maintaining higher levels for longer, earning the name the 'warrior gene'.

Similar studies conducted in Sweden and New Zealand focusing upon the extremes of antisocial behaviour also support the notion that genes have a role to play in criminal activity – however, they are all small and so statistically under-powered to show anything conclusive. As yet, there isn't enough worldwide evidence to support any notion of genetic testing as a means to predict criminality in a scientific manner.

▼ Some nerve cells, or neurons, produce a chemical called dopamine, a neurotransmitter that controls nerve impulses important in movement, emotion, pleasure and pain.

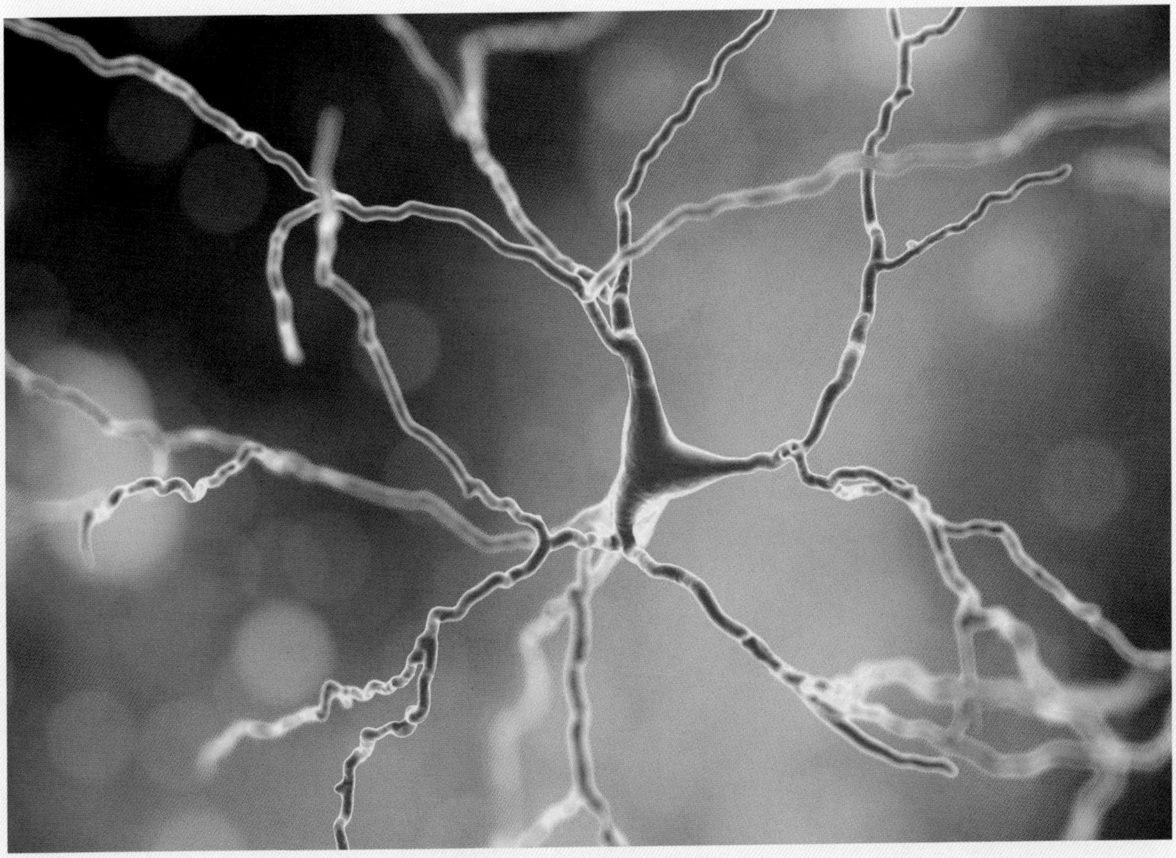

DNA fingerprinting

The first foray of DNA profiling in forensics came when Leicestershire police were convinced a double killer was responsible for the rape and death by strangulation of two 15-year-old girls, based on the similarities between crimes committed two-and-a-half years apart. They carried out a 'DNA sweep' to gain as many samples from men in the local area as possible. They compared these DNA fingerprints to DNA extracted from samples found at the crime scenes. However, their efforts were fruitless – none matched. A year later, a man overheard someone in a pub bragging that, during the sweep, he had persuaded an associate to provide a sample in his place. Police then tested the bragging man's DNA and the profile proved to be a match. The man was convicted in the first case brought to justice using DNA fingerprinting.

Since then, estimates are that investigators over the world have taken the DNA profiles of some 50 million people. DNA profiling has played a key role in the identification of millions of criminals – and even a long-dead king.

The king in the car park

In August 2012, a team from the University of Leicester began searching for the remains of King Richard III (1452–1485). Richard was King of England and Lord of Ireland from 1483, until he died at Bosworth Field, a decisive battle in the Wars of the Roses. His death at the hands of Henry Tudor, who was later to become King Henry VII, ended the reign of the House of York. After his death in the field, Richard's body was taken to nearby Leicester and buried without ceremony in the Church of the Grey Friars in a tomb that was misplaced during the events of the Protestant Reformation in the 16th century. Nevertheless the archaeological team, hoping to find Richard's grave, hedged their bets that he was still within the Grey Friars.

Historical records showed that the site was now underneath a car park and the team, who were working on a very tight budget, had to make a crucial decision as to where the search should begin. They believed his tomb would have been placed in the choir of the church. They began digging – and found skeletal remains. But did they belong to King Richard III? The circumstantial evidence was in their favour: the remains were in the right place, they belonged to a man aged 30–34 years old and he had multiple battle injuries. Radiocarbon dating showed that he had died some time between 1450 and 1540. There was evidence of scoliosis,

▼ Lead archaeologist Richard Buckley, at the University of Leicester reveals whether the human remains found in Leicester are those of King Richard III. (Getty)

A true royal likeness

Identifying Richard's remains was not the end of the story for the DNA that the scientists recovered from his skeleton. The Leicester team had another use for it: they wanted to use it to find out what he looked like. In particular, the researchers wanted to know the colour of Richard's hair and eyes. There are two known portraits of Richard, one in the Royal Collection and another in the Society of Antiquaries of London. But which (if either) provides the closest likeness? Looking at genetic markers linked with these physical traits showed with 96% probability that Richard had blue eyes and with 77% certainty that he was blonde (at least as a child). It seemed that he most closely resembled the portrait held by the Society of Antiquaries.

curvature of the spine, which Richard was known to have had. So far, so good, but none of this was definitive proof that these were actually the remains of Richard III.

Scientists successfully extracted DNA from the bones and so began the process to show genetically that this was indeed the king. First, the focus was on Y-chromosome DNA. Richard himself did not have any direct descendants, so unravelling the mystery required first going back a few generations, then setting off down an alternative branch of the family to find living relatives. In a feat of genealogical cunning, the trail went back to King Edward III (1312–1377) and John of Gaunt (1340–1398), then jumped a number of generations to the Duke of Beaufort (1744–1803), before finally making a link to the present-day Somersets, Richard's male-line descendants. But it stopped here: a completely discrepant Y-chromosome pattern to Richard's own revealed a non-paternity event somewhere along the line.

The DNA detective work focused on the maternal line instead. Richard's mother, Cecily Neville, will have given her mitochondrial DNA to him and his sister, Anne of York. Anne, in turn, would have passed hers on to the next generation. In fact, it was possible to follow a trail down the maternal line and ultimately to two people alive today – Michael Ibsen, a 17th-generation descendant, and Wendy Duldig, a 19th-generation relative. Comparing their mitochondrial DNA with that from the remains found in the Leicester car park showed a complete match with Ibsen's and only a single base difference with Duldig's. The results confirmed that the remains were, indeed, King Richard III. The mystery solved through DNA testing, Richard's body was reinterred, this time with ceremonial pomp and circumstance, on 26 March 2015, in Leicester Cathedral.

▼ Michael Ibsen (L) and Wendy Duldig, descendants of England's King Richard III, address a press conference at Leicester Cathedral in Leicestershire on 23 March 2015, ahead of the king's reburial some 530 years on from a violent death in 1485 at the Battle of Bosworth. (Getty)

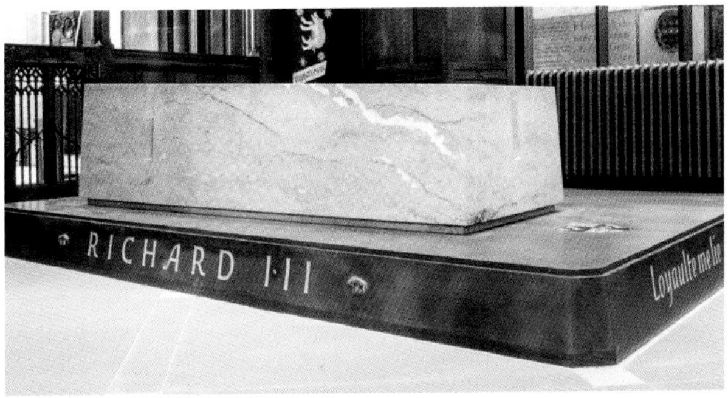

The O.J. Simpson Trial (1995)

On 17 June 1994, 95 million people tuned in to watch O.J. Simpson, US NFL football star turned TV sports pundit, being pursued by police. He had failed to turn up at police headquarters that morning to answer questions about evidence that implicated him in the murders of his ex-wife, Nicole Brown Simpson, and her friend, Ronald Goldman. He had been permitted to attend her funeral on the understanding he would hand himself in the next day. He was caught, apprehended and put on trial in televised court proceedings that would last 133 days. When the not-guilty verdict was finally returned, the world watched on in utter fascination.

DNA evidence, in its infancy as a forensic tool, was to play a role in the jury's decision. However, before the trial, officials discussed whether it could be admissible in court; after all, its significance was still little understood and its validity shrouded in suspicion. Nonetheless, the prosecution presented the jury with the results of DNA fingerprinting gathered from a number of different sources at crime scene and from Simpson's home. Analysis showed that blood found at the scene had a 1 in 170 million match with Simpson's and blood found at this home had a 1 in 6.8 billion match with his murdered ex-wife. By today's standards the evidence was compelling – but the defence team simply batted it away, stating, 'Something [must be] wrong.' Simpson walked free.

▼ Double murder defendant O.J. Simpson (C) sits with his attorneys Johnnie Cochran Jr (R) and Robert Blasier (L) during a court hearing in the O.J. Simpson murder trial 11 September 1995. (Getty)

A brush with DNA

Is it possible to extract DNA from a hairbrush or toothbrush? Yes, it is.

The source of DNA in a hair is from the follicle, the sheath of cells that covers the hair root, anchoring it to the dermis layer of the skin. (The hair shaft itself does not contain DNA, as it has no cells, just a protein called keratin.) So, the hairbrush would need to contain hairs that have been plucked out from the scalp with the root intact. Toothbrushes provide a source of DNA because they are coated in saliva, which contains cheek cells.

Members of the original prosecution team are convinced that were they to conduct that trial, using the same evidence, again, the outcome for Simpson would be very different. Nowadays, such is our faith in DNA evidence that the judicial system regularly relies upon it. We protect our crime scenes as quickly and as efficiently as possible to avoid sample contamination and to limit conflicting results from the analysis. And what self-respecting TV criminal investigation show would be complete without the DNA forensics team?

Paternity testing

Ever since Alec Jeffreys' discovery that our inherited DNA is identifiably from our parents, but that our DNA fingerprints are unique, there has been a demand to apply this knowledge to establishing correct parentage. In the past, we could do this only using blood samples from both parents and the child. Nowadays, though, because of technological advances, we can carry out paternity testing using saliva or even hair follicles. Furthermore, we need only samples from the putative father and the child – so-called 'motherless' testing. Scientists analyse 35 different minisatellite markers and compare the numbers of repeats in the child's sample with the adult's. A match with the father confirms paternity, whereas differences indicate non-paternity.

DNA and victims of disaster

Natural – or other – disaster, leading to multiple victims, is one of the most societally important uses for DNA profiling. Identifying victims correctly in order to bring families closure and to enable loved ones to be laid to rest is possible if we can match the genetic fingerprints in DNA extracted from remains with either DNA from the victim (using their toothbrush as a source, for example), or with the DNA fingerprint of a relative.

▲ **Search and rescue teams look for people who were in a building when it collapsed.**

Coordinating a response to a mass disaster using DNA profiling to identify the victims is a major logistical exercise. Creating a system to collect and catalogue samples, to analyse the DNA fingerprints and to compare them with other DNA profiles involves a meticulous methodology and painstaking attention to detail, often in the face of the most inhospitable working conditions. Fire, hurricane, explosion, tsunami – and myriad other disaster scenarios – can spread debris far and wide, fragment bodies and make recovery incredibly difficult. Furthermore, matching the DNA profile between victim and sample relies upon help from relatives. Scientists need loved ones to come forward in the face of their heartache to offer their own DNA sample or a sample from a potential victim's hairbrush or toothbrush, say, in order for profiling to offer up any answers.

It's hardly surprising, then, that the process of using DNA to identify victims can take some time. Following the 2018 wildfires that ripped through northern California and devastated the town of Paradise, officials relied initially upon fingerprints and dental records to identify the victims. For the remains that were very badly damaged, though, positive identification rested upon DNA profiling. However, the relatives' DNA database remained under-populated – three weeks after the fires, 88 people had been identified, but up to 200 were still listed as missing.

From London's Grenfell Tower fire in 2017 to the South East Asian tsunami in 2004 and the Ethiopian Airlines plane crash in 2019, DNA profiling has been the mainstay of returning the bodies of loved ones to their relatives in the face of complete devastation. But DNA is wonderful and long-lasting and even decades after the event it can provide closure for those who need it: DNA analysis to identify victims of the 2001 terrorist attacks on New York's World Trade Center is still ongoing – and successfully identifying victims, almost two decades after the tragedy.

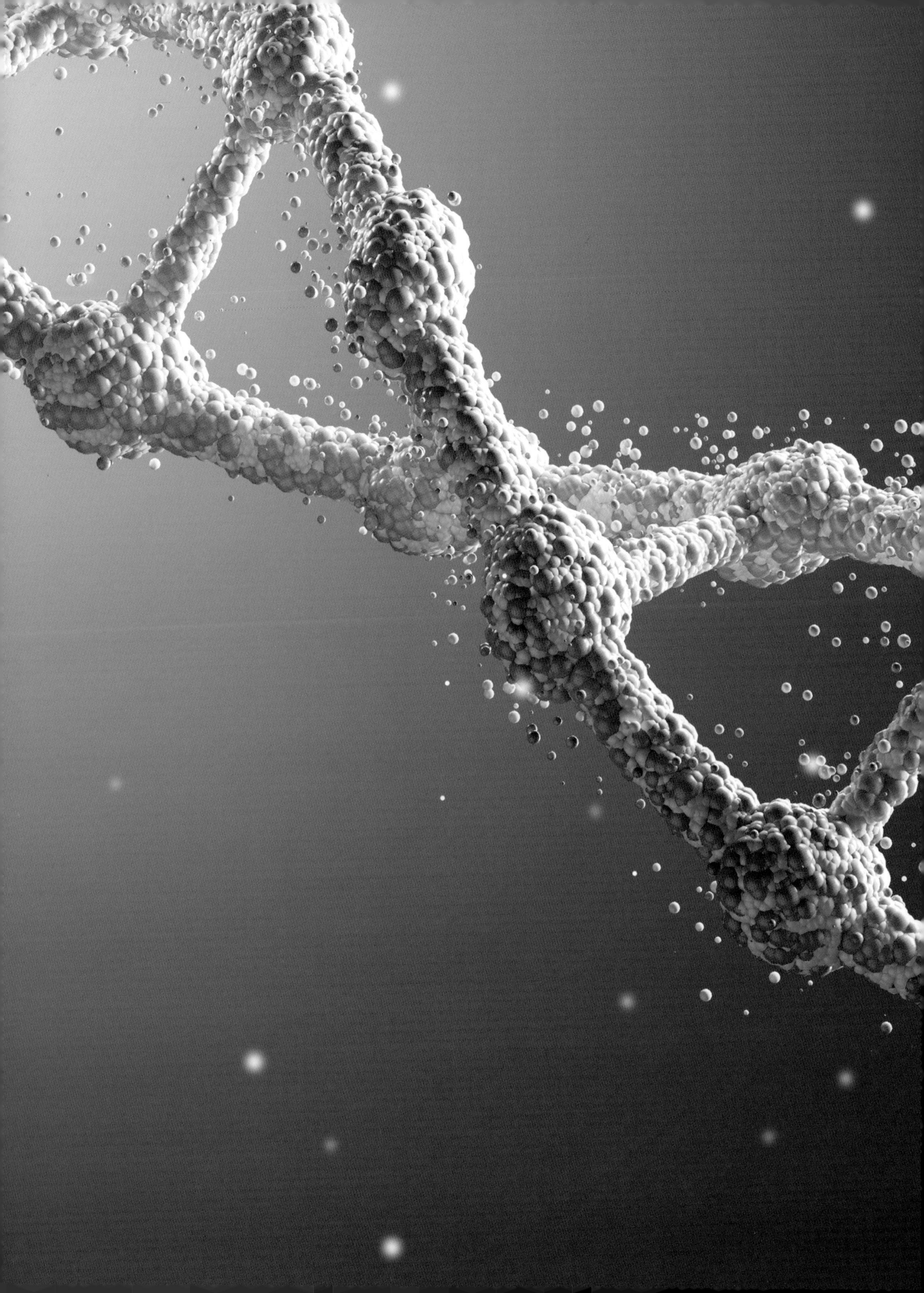

CHAPTER 6

DNA in medicine

Ever since scientists completed the task of mapping the human genome in 2001, we have had a much better understanding of the role our DNA plays in causing genetic diseases. Add to this recent technological advances that mean we can readily access the genome and we have a genetics health revolution on our hands.

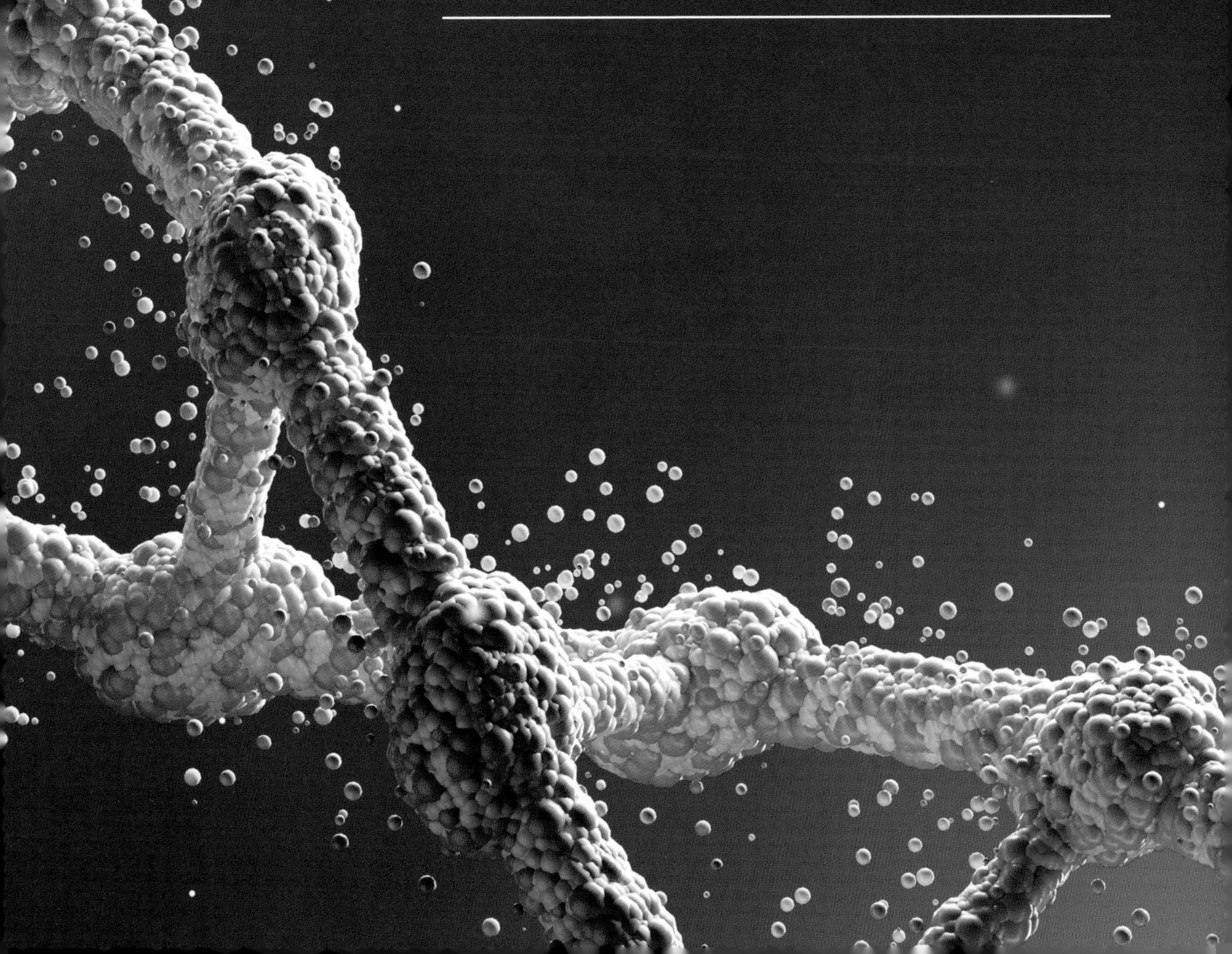

DNA & the health revolution

We have to go back to Mendel and his principles of inheritance to understand genetic risk and what we can do to modify it.

Patterns of inheritance give an indication of how a genetic condition is passed down through a family. They can be defined by the location of the gene that is responsible for the trait – in other words, chromosomal or mitochondrial. If a trait is passed on through a chromosomal gene, its inheritance may be autosomal or dominant or recessive, or sex linked. Let's start with dominant.

▼ Technological advances in genetic testing have revolutionised the use of DNA in medicine today.

Autosomal dominant

An autosomal dominant (AD) condition is caused by a pathogenic variant on one copy of a pair of genes. The variant dominates over the other copy, known as the wild type (WT). Distinctive characteristics of an autosomal dominant inheritance pattern are that males and females can both be affected with equal probability, males can transmit to other males from one generation to the next, the genetic variant associated with the trait is passed on with 50:50 chance, and it can show itself in more than one generation.

Dominant genetic conditions can also be *de novo*, meaning it is a new event that comes out of the blue, it is not inherited from either parent and there is no family

Autosomal dominant inheritance

The key characteristics of an autosomal dominant trait are that it can affect both male and females, it is passed down from generation to generation with 50% probability and there can be male to male transmission.

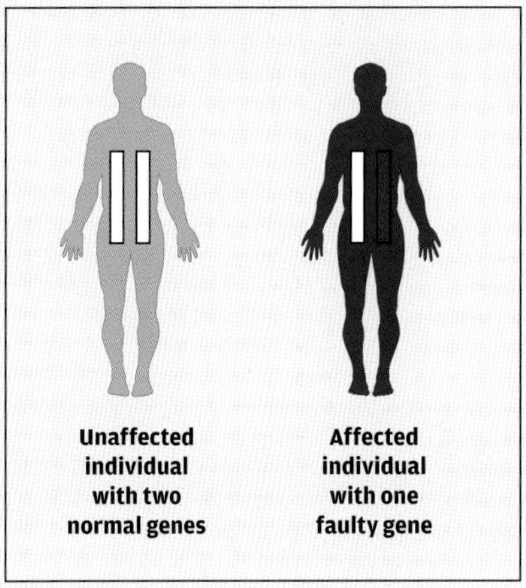

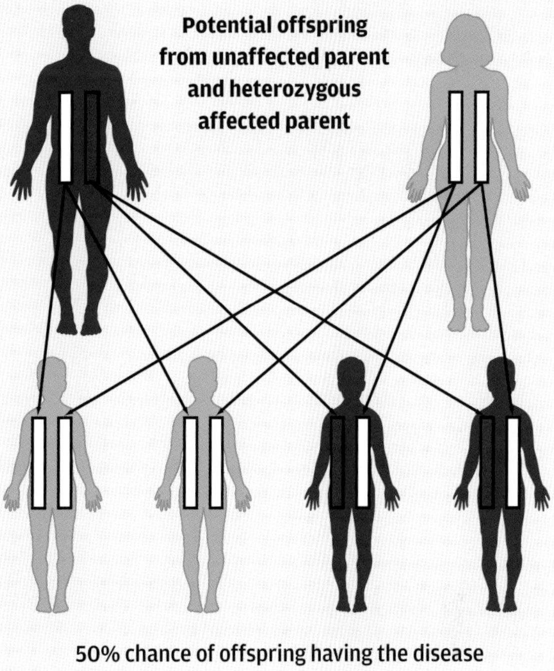

history of it. This, too, can be passed on with a 50:50 chance. New dominant changes are more likely when the father is older, in the same way that there is an increased chance of chromosome defects in the babies of pregnant older mothers.

Neurofibromatosis type 1

Neurofibromatosis type 1 (NF1) is an example of an AD genetic condition that affects the skin and the nervous system, known as a neurocutaneous disorder. It is caused by a whole host of pathogenic variants in the gene that makes a protein called neurofibribromin, NF1, which is found in nerve cells throughout the body. NF1 is truly representative of an AD genetic syndrome in that it displays many characteristic features. What gives the condition its name are the benign swellings (tumours) on the sheaths around the nerve fibres called neurofibromas. They appear as lumps in the skin, which may also have multiple areas of increased pigmentation called café-au-lait patches, as they resemble splashes of coffee, and freckles in the armpits and groin. Little nodules in the iris, the coloured part of the eye, might develop and, while they do not cause any problems with vision, these nodules appear on the diagnostic checklist for NF1. There is a relatively high *de novo* rate in NF1, so not everyone with it has a suggestive family history.

Some people with NF1 syndrome have epilepsy and intellectual disability, and are more likely to develop certain cancers, such as breast cancer and tumours of the optic nerve – the major nerve that sends messages from the back of the eye to the brain. For this reason, if you have NF1 you need to have regular, lifelong check-ups to detect any complication as soon as possible.

Another interesting genetic phenomenon of NF1 is germ-line mosaicism. This is the reason why NF1 sometimes recurs in siblings, even if neither parent is affected. In this situation, NF1 was not truly *de novo* in their child – it was not really the result of a random change that occurred in the NF1 gene of the single egg or sperm. Instead, the NF1-causing variant was present all along in the germ-line cells from which the egg or sperm cells were made. And this means they can pass on the genetic change again, making the chance of recurrence not zero, but more like 5–10%. Sometimes NF1 can display mosaicism in other tissues, too, such as the skin, where there can be just one strip of skin on the body with the characteristic pigmentation markings. This signifies the gene change is present only in some places in the body.

Genetic counselling in NF1 can be very complex because of the unpredictable variability in how it affects people and because of the different kinds of mosaicism that might come into play – it is not usually possible to determine if NF1 occurred in someone as a truly new event or if it was because of the gene-line, which only really becomes apparent if it happens again.

Autosomal recessive

A trait or genetic condition is autosomal recessive (AR) in its inheritance when two parents are carriers and they both pass on the pathogenic variant to their offspring, who then gets a double dose of the faulty gene. Each parent has a 50:50 chance of passing on the variant, which means there is a probability of one in four that together they will have a child with the AR condition each time they conceive together. In a typical AR inheritance pattern, only one generation is affected, males and females with equal probability, and there might even be consanguinity where couples who are closely related, like first cousins, have children together (something that is relatively common in some ethnic groups). In this case, both partners are more likely to carry the same damaging genetic variant.

We have already come across the concept of people from some ethnic backgrounds being more likely to carry certain recessive genetic changes, such as cystic fibrosis (CF) in Caucasians and sickle cell disease in Africans. Genetic screening for recessive conditions is sometimes offered on the basis of the ethnic background you come from.

Cystic fibrosis

Cystic fibrosis (CF), an AR genetic condition, is a serious multisystem disease that affects breathing and digestion because of its effects on the lungs and pancreas. People with CF have thick and sticky mucus, which can clog up the lungs and the part of the pancreas that produces amylase, a food-digesting enzyme. This causes repeated chest infections and problems associated with nutrition and growth from early childhood. Over time, the lungs become scarred through repeated infections, a process called fibrosis. While we can treat, or even prevent chest infections, and we can add artificial digestive enzymes to food, CF is steadily progressive and it is not currently curable.

One in every 25 Caucasians are CF carriers (people from other ethnic backgrounds can also carry CF, but it is less frequent), and couples who are both carriers have a one in four chance of having affected children each time they conceive together. This makes the incidence around one in every 2,500 births, so it is one of the most common genetic conditions, to the point that babies are tested for it as part of the newborn screening (heel prick) tests. The tests also look for eight other recessive conditions, including some inborn errors of metabolism, sickle cell disease and deficiency of the thyroid gland hormone, thyroxine, all of which can be successfully treated if they are detected early.

A number of different pathogenic variants are known to cause CF, but there is one that is much more common in the *CFTR* (cystic fibrosis transmembrane conductance regulator) gene called F508del – which causes the deletion of the phenylalanine amino acid at position 508 in the CFTR protein. This protein forms a channel, sitting across the cell membrane

Autosomal recessive inheritance

The key characteristics of autosomal recessive inheritance are that both parents are usually healthy carriers, both males and females can be affected, and it tends to be found in one generation.

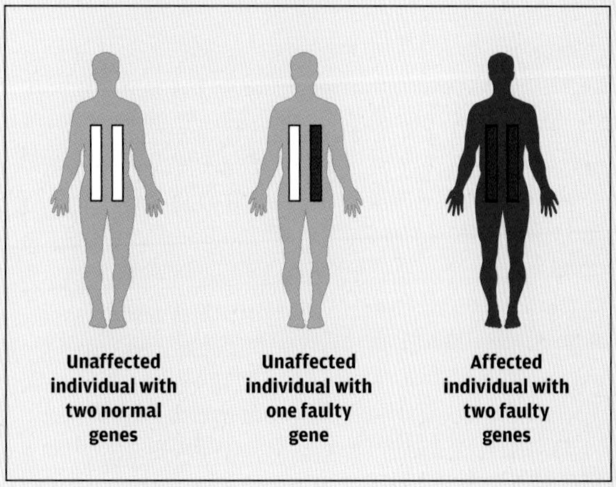

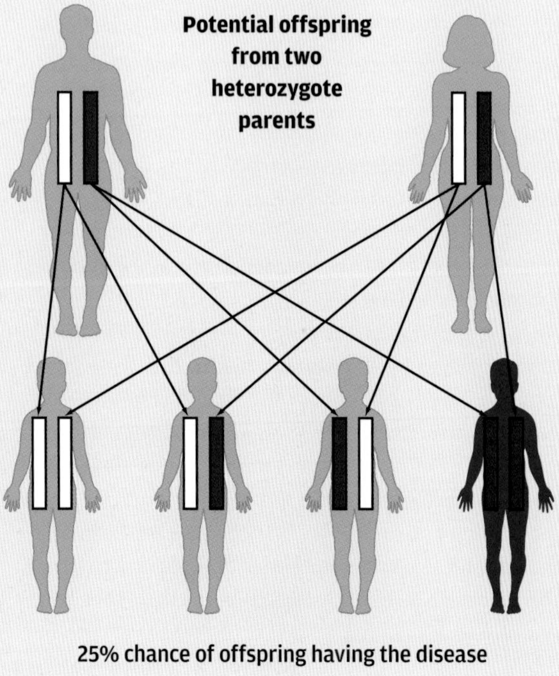

Cystic fibrosis

Cystic fibrosis is an autosomal recessive condition in which the mucus the body produces is too thick and sticky, the effect of the CFTR gene change on the salt content. The mucus clogs the airways of the lungs and the channels of the pancreas, stopping them working properly and damaging the surrounding tissue.

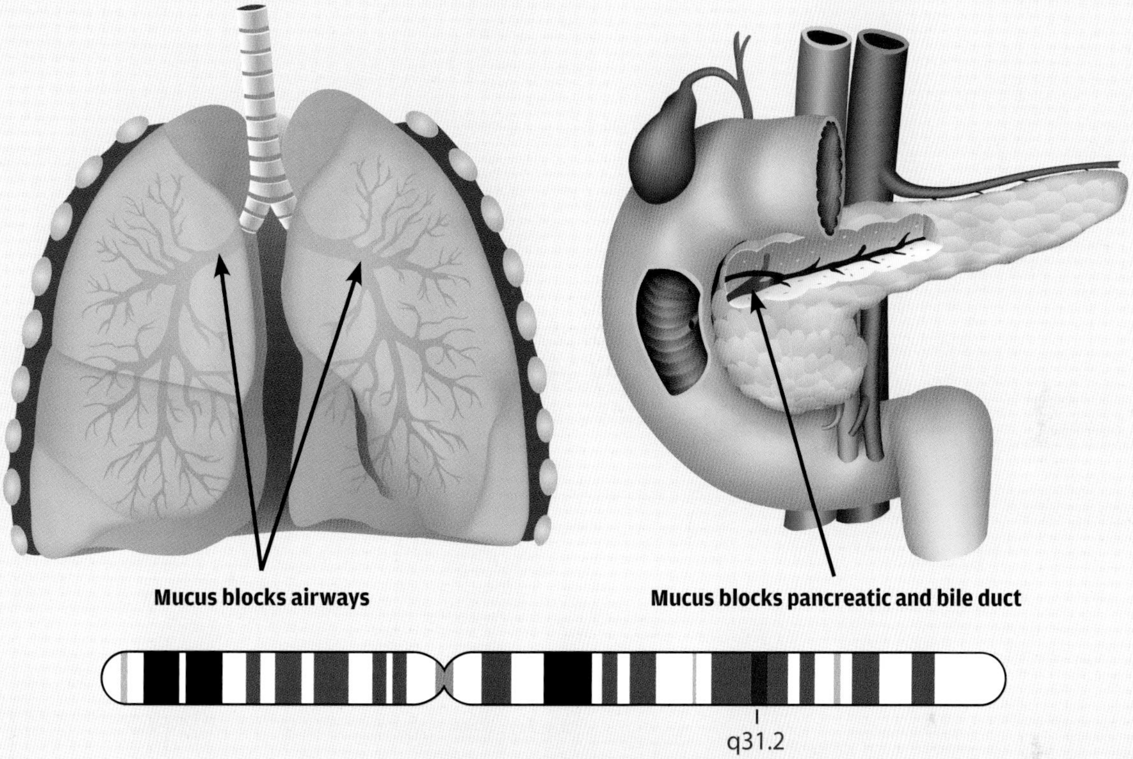

The location of the CTFR Gene on chromosome 7
(CFTR = Cystic fibrosis Transmembrane Conductance Regulator)

in respiratory cells in the lungs, moving salts in and out. The channel malfunctions in CF and the altered salt content of the mucus makes it far too thick and sticky. Understanding the effects on CFTR of the different, damaging variants means that we might one day be able to develop new treatments using gene therapy and gene modifiers that might much more effectively treat this relatively common, serious genetic condition – and perhaps even prevent it occurring altogether.

X-linked inheritance

X-linked (XL) inheritance occurs when one of the thousand or so genes on the X chromosome is altered. Females have two X chromosomes and males have one (their other sex chromosome is the Y). With the exception of a small number of genes in the pseudo-autosomal regions (PAR) of the sex chromosomes, there are different genes on the X and the Y.

Consanguinity

If you have children with someone from the same family as you, there is an increased probability that you share a disease-causing variant inherited from a common ancestor, such as a grandparent. This is called consanguinity.

There are different degrees of consanguinity depending on the degree of relatedness of the parents – the further apart they are in the family tree, the less likely it is that it will cause a problem. It is not inevitable that couples who are related will have children with a genetic disorder, but it is more likely than it is for couples who are unrelated.

X-linked recessive inheritance

In contrast to autosomal dominant inheritance, the key characteristics of X-linked recessive inheritance are that males are affected, females are usually healthy carriers, and there is no male to male transmission. Multiple generations can be affected.

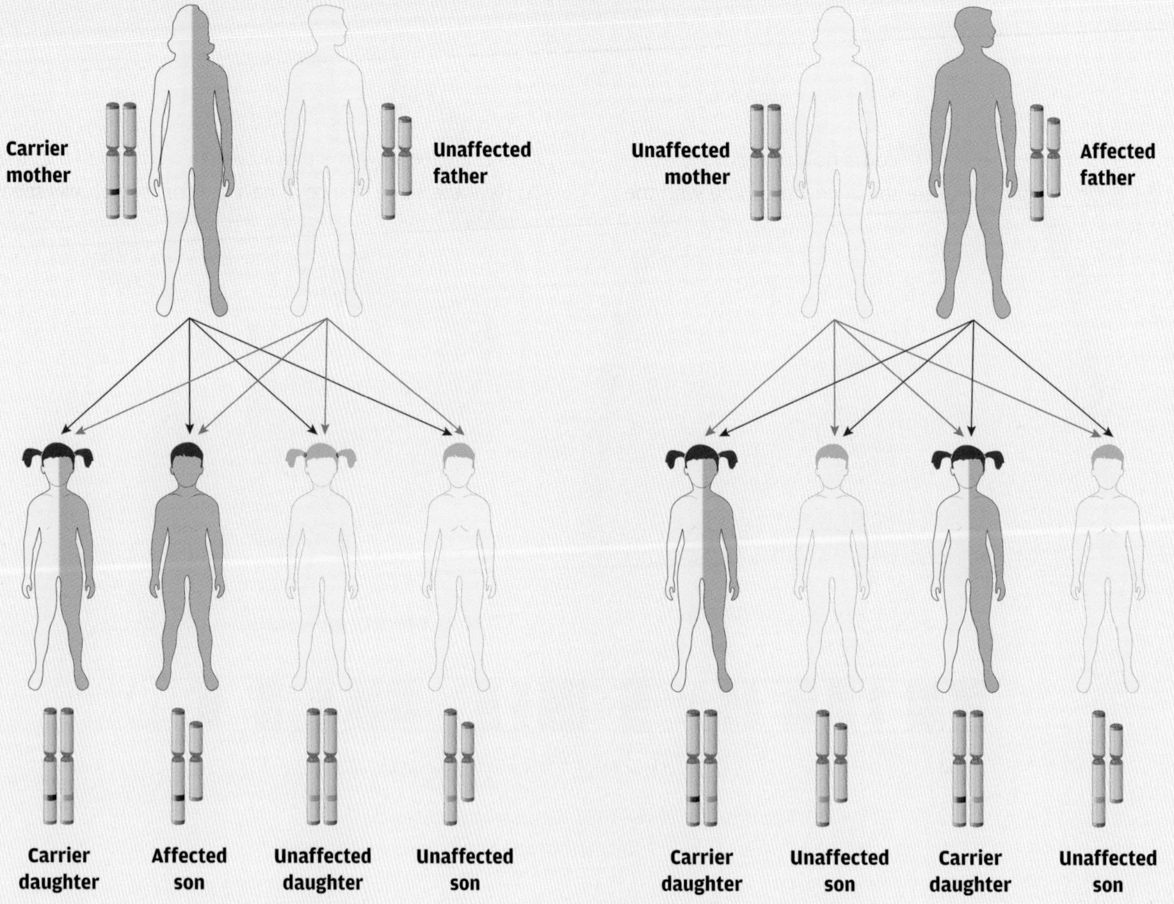

X-linked recessive inheritance (XLR)

Females who carry an X-linked recessive (XLR) disorder have one faulty X-chromosomal gene copy, but they are healthy because they have a second copy to compensate. Males, on the other hand, do not – which means a faulty X-chromosomal gene copy will certainly cause an XLR condition. The typical features of XLR inheritance patterns are that it is possible for males to be affected across many generations, but one male can never pass it on to another – they are male because they inherit the Y-chromosome from their father, not the X, so the link between one affected male and another is only ever through female members of the family. (There are some exceptions to the rule that females are not affected by an XLR genetic condition, where carriers may have a much milder form.)

X-linked dominant inheritance

X-linked dominant (XLD) inheritance causes a disorder that also affects females, because the faulty gene dominates over the other, good copy. It often follows, though, that the condition is much more severe in males. Consequently, affected male foetuses might not survive the pregnancy. The exception to the rule is that sometimes males and females can be equally affected with a significant genetic condition, but one that is not life-limiting. For example, X-linked hypophosphatemic rickets is a genetic form of rickets not caused by lack of vitamin D, but instead by a defect in an XL gene that means the body cannot absorb the calcium and phosphate salts it needs to make the bones strong and straight. Although it is an X-linked condition, it affects both sexes equally.

X-inactivation

The process that happens in females to 'switch off' or inactivate one of their X-chromosomes – remember males need only one copy – by lyonisation is usually random. It is either the paternally inherited X in a female's cells that gets switched off or it is the copy that came from the mother. There is no preference to switch off one copy over the other. (The result of the same random process in female tortoiseshell cats gives them their distinctive coat colours.) However, in conditions where a genetic defect involving an XL gene occurs, the process can become non-random – that is, females preferentially inactivate the X-chromosome with the faulty gene copy. That is another reason why females are less likely to show physical traits associated with an XL disorder.

Duchenne muscular dystrophy

Duchenne muscular dystrophy (DMD) is a progressive muscle disorder that mostly affects boys because it is X-linked. The gene, the largest known in humans, which becomes altered to cause DMD, is responsible for making a protein called dystrophin. This is the protein that makes skeletal muscle and heart muscle strong. Without it, muscles are much weaker, so that boys with DMD who have abnormal dystrophin learn to walk much later than other children and in general lack normal physical strength. It is a progressive condition that means the muscles become steadily weaker as boys get older and eventually, usually in their teens, affected boys are no longer able to walk.

This is a life-limiting condition with no cure, although scientists are doing a lot of research to find a gene-therapy treatment and ways to overcome the genetic defect.

Being a carrier of DMD has much less of an impact on females, who are generally well. Sometimes, though, a woman might have a mild weakness in her skeletal or heart muscle. This is because X-inactivation has happened in a non-random way in her cells, and more of the good than the bad copies have been switched off.

Female carriers have a 50:50 chance of passing on the DMD change every time they have children. Given that they also have the same chance of having a son, statistically speaking half the number of their sons will be affected and half the number of their daughters will be carriers. (All the daughters of an affected male will be carriers and none of his sons will be affected, but in reality affected males rarely have children given the severity of DMD.)

There is a high rate of *de novo* variants in DMD, but where it runs in the family, successful use of prenatal testing and other reproductive technologies have been transformative.

Y-linked inheritance

Y-linked (YL) inheritance occurs when a father passes genes on his Y-chromosome to his sons, so only male-to-male transmission can occur. In reality, we hardly ever see YL inheritance because the genes on the Y chromosome are responsible for male sexual development and where these are faulty, infertility is the rule, making transmission impossible.

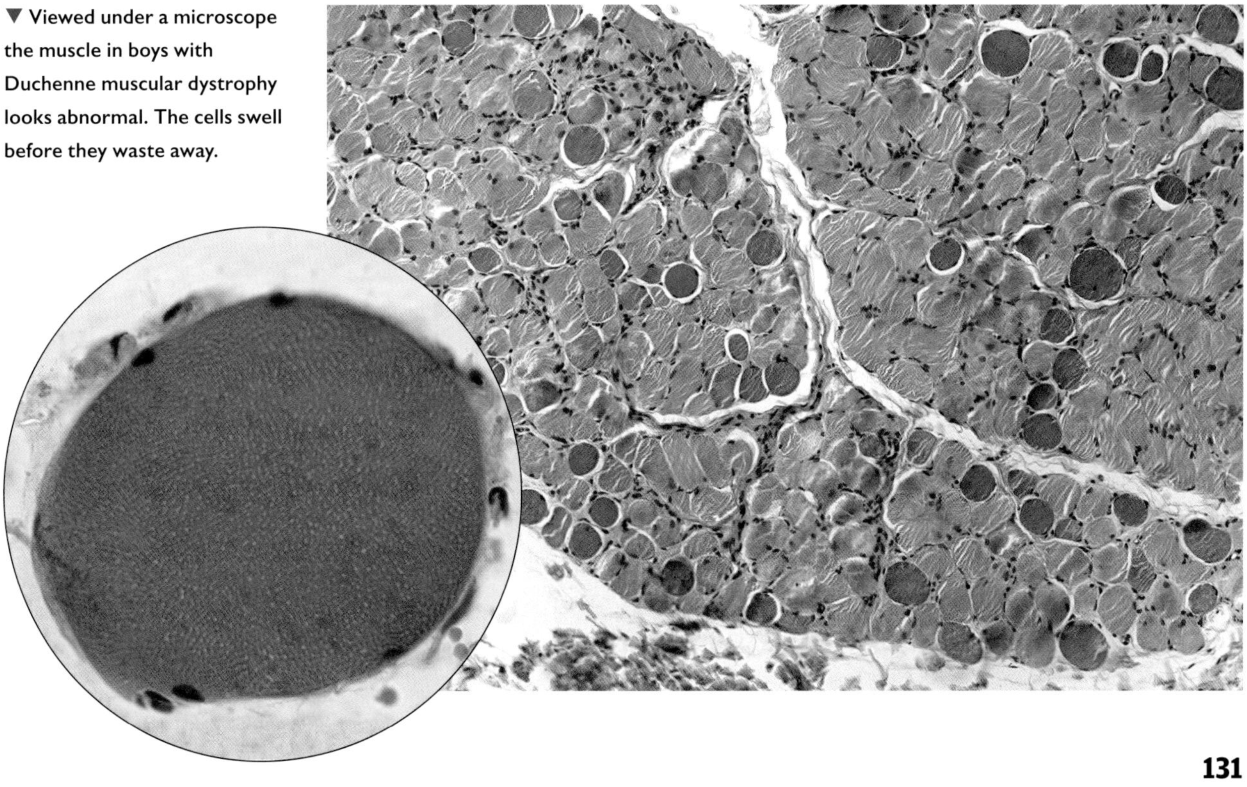

▼ Viewed under a microscope the muscle in boys with Duchenne muscular dystrophy looks abnormal. The cells swell before they waste away.

Royalty, Rasputin and the Romanovs

Haemophilia A is a disorder of the blood's clotting system, which is meant to stop bleeding. The body produces a cascade production of many proteins in response to cut blood vessels. An important component of the clotting cascade is called factor VIII, a protein made by an X-linked gene, *F8*. Females have two *F8* copies and males only one. Therefore pathogenic variants in *F8* result in severe deficiency of this crucial clotting factor in males, which is known as haemophilia, a clotting disorder in which there is a tendency to bleed catastrophically, even after a minor injury. It brings with it significant risk of health complications, such as arthritis, from bleeding into joints, and even death from internal bleeding. Today, haemophilia A is treated by replacing factor VIII directly through regular intravenous infusions and through gene therapy.

Back in 1900s Russia, though, there was no such option, except for, in the case of the Tsarevich Alexis Nikolaivich, 'treatment' through hypnosis administered by the self-proclaimed holy man, Gregorii Rasputin. Alexis was a member of the Romanov Imperial Family, the son of Tsar Nicholas II of Russia. His mother, Alexandra (Alix), the Tsarina, was Queen Victoria's granddaughter, the sixth-born child to her daughter, Princess Alice. Alice had a son, Frederick, who had haemophilia A, just like Queen Victoria's son, Leopold, illustrating in the Royal Family a typical picture of the X-linked inheritance pattern of this condition. All of this meant that Alix had a 50:50 chance of being a carrier of haemophilia A. Before giving birth to Alexei, she had three daughters, but when she had her first son, it very quickly became obvious to her that he had the family's clotting affliction, because bleeding from his umbilical stump would not stop.

Long before factor VIII was even discovered, and many decades before replacement therapy was developed, Rasputin offered to 'treat' Alexei's condition, using hypnosis to sooth him out of his excruciating haemophilia crises. Rasputin's demise is legendary – he was poisoned, shot and drowned. So was Alexei's. He was executed along with his Romanov family in 1918, during the Russian Revolution, their bodies thrown down a mineshaft before later being retrieved to be burned.

The remains of the executed Imperial Royal Family were recovered in 1979 and underwent genetic identification using DNA fingerprinting and mitochondrial DNA analysis, one of the reference samples used for comparison being that of Prince Phillip, the Duke of Edinburgh (who is the great-nephew of Alexandra, Alexei's mother). (In a related plot twist, DNA technology was called upon again to solve a mystery. Alexei's younger sister, Anastasia, who was missing from the site where

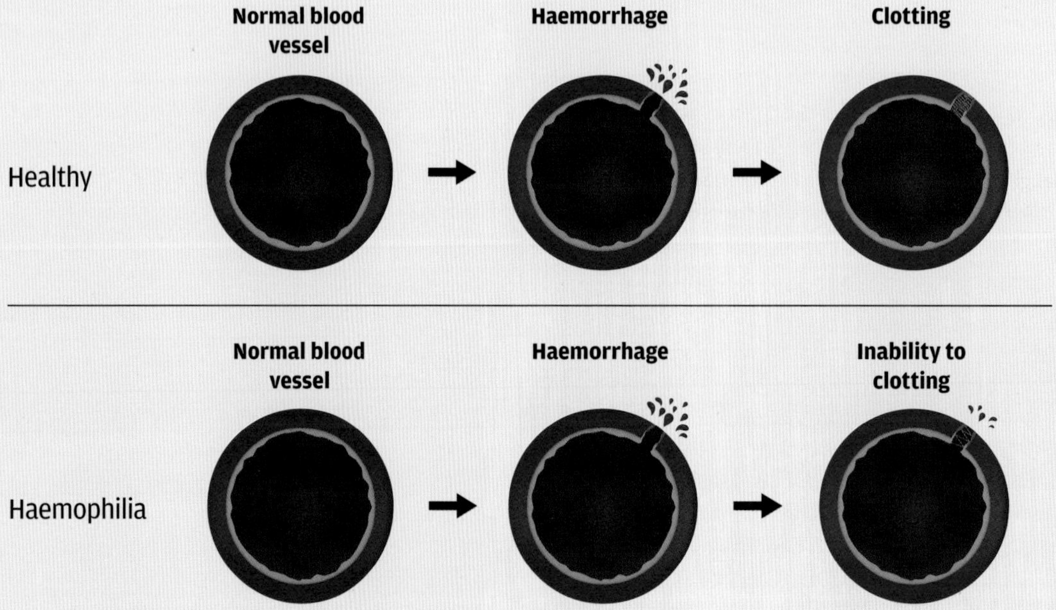

▶ Cousins Nicholas (left) and George (right), later to become Tsar of Russia and King of England respectively, shared a distinct family resemblance, but fortunately for the Windsor family, George escaped inheriting the faulty *F8* gene.

the remains of her family were discovered, seemingly resurfaced in the form of Anna Anderson, a Polish inmate at a mental asylum in 1920s Berlin. Her lifelong claim was ultimately disproved when, after her death in 1984, DNA fingerprinting revealed she was not Anastasia.)

Haemophilia A persisted in this family through a different line of descent from Queen Victoria. Her youngest daughter, Beatrice, had two sons affected by haemophilia A and her own daughter, Victoria Eugenie, turned out to be a carrier as well. After Beatrice married Alphonso XIII of Spain in 1906, she had four sons, two of whom had haemophilia A and both of whom died. Her fourth son, Juan, unaffected by the familial disease, became King of Spain with no further transmission of the condition through members of the Spanish Royal Family.

What does this family history mean for the current British Royal Family? Might it resurface? Has it 'skipped generations'? In fact, it is unlikely to have any consequences at all. In a brief genetic counselling overview, here is how they are connected to their ancestors who must have been carriers:

Queen Elizabeth II is distantly related to her husband, Prince Phillip – her great-grandfather, Edward VII, and Prince Phillip's great-grandmother, Alice (mother of the Tsarina, Alix), were brother and sister, and were two of Queen Victoria's children. Queen Victoria must have been a carrier, as she had an affected son, Leopold, as well as daughters who themselves had affected sons. To link her with Queen Elizabeth II, we pass through male relatives in three subsequent generations: Edward VII, Victoria's son; his son, George V; and his son, George VI – Queen Elizabeth's father. As George VI was unaffected and this X-linked condition has to pass through the female line, the Queen cannot have inherited the haemophilia A-causing genetic variant. The Duke of Edinburgh is descended from three female relatives who can be traced back to connect his bloodline with Queen Victoria. But he himself does not have haemophilia, which means that the faulty *F8* gene has not been transmitted along the female line to him. Either way, the condition has therefore disappeared from the British Royal Family.

Mitochondrial inheritance

When the mitochondria (singular, mitochondrion), those thousands of tiny cell organelles found within the cytoplasm, do not function properly, a mitochondrial disorder occurs. Because the mitochondria are the powerhouses of the cell, tissues with the highest energy requirements, such as the brain, the liver and the muscles, suffer the impact the most. Mitochondrial disorders can happen for one of two reasons: there is a genetic fault in the nuclear DNA or in the DNA contained inside the mitochondria (mtDNA).

To function efficiently, mitochondria rely on proteins produced by genes located in the nuclear DNA. Sometimes the reason for a mitochondrial disorder is because the body is not producing one of these proteins properly, a situation that usually follows an autosomal recessive inheritance. When both parents are carriers, there is a one in four (25%) chance of them both passing on the genetic change and having a child with a mitochondrial disorder.

The situation becomes rather more complex when a mitochondrial disorder is the result of a pathogenic variant in mtDNA. Variants in mtDNA can be deletions or duplications involving a whole run of nucleotide bases or they can just involve a single base. Deletions and duplications are usually *de novo*, a new change in the mtDNA, not an inherited one, and the chance of this happening again in siblings of an affected child is very low.

Single base changes can be either *de novo* or inherited, which one it is depends on whether the mother is affected. As nearly all our mitochondria originate from the egg, if the mother is not affected by a mtDNA mitochondrial disorder, there is a much lower chance that it would recur – but it is not impossible: in practice, the chance is one in 24. If the mother is affected though, the recurrence can be as high as 100%.

A key characteristic of mitochondrial inheritance is transmission through the female line, because mitochondria are inherited almost exclusively from the mother via her egg (only a tiny number come from the father in his sperm). In distinction to XL inheritance, males cannot pass on the mtDNA variant because it stops with them and both sexes are equally as affected. Many generations can be affected by mitochondrial inheritance, which is different to AR inheritance, where it does not involve any subsequent generations.

Heteroplasmy

Because one cell contains thousands of mitochondria, it is possible for the mtDNA in some mitochondria to have a pathogenic variant, while others do not. It is also possible that the mtDNA in one group of cells contain a pathogenic variant that is not present in another group, so that the level of pathogenic mtDNA to normal mtDNA can differ from tissue to tissue. The presence of different mtDNA populations is called heteroplasmy.

Mitochondrion

The mitochodria are the powerhouses of the cell. There are thousands of these organelles within the cytoplasm, more in organs and tissues that have higher energy requirements, like the brain.

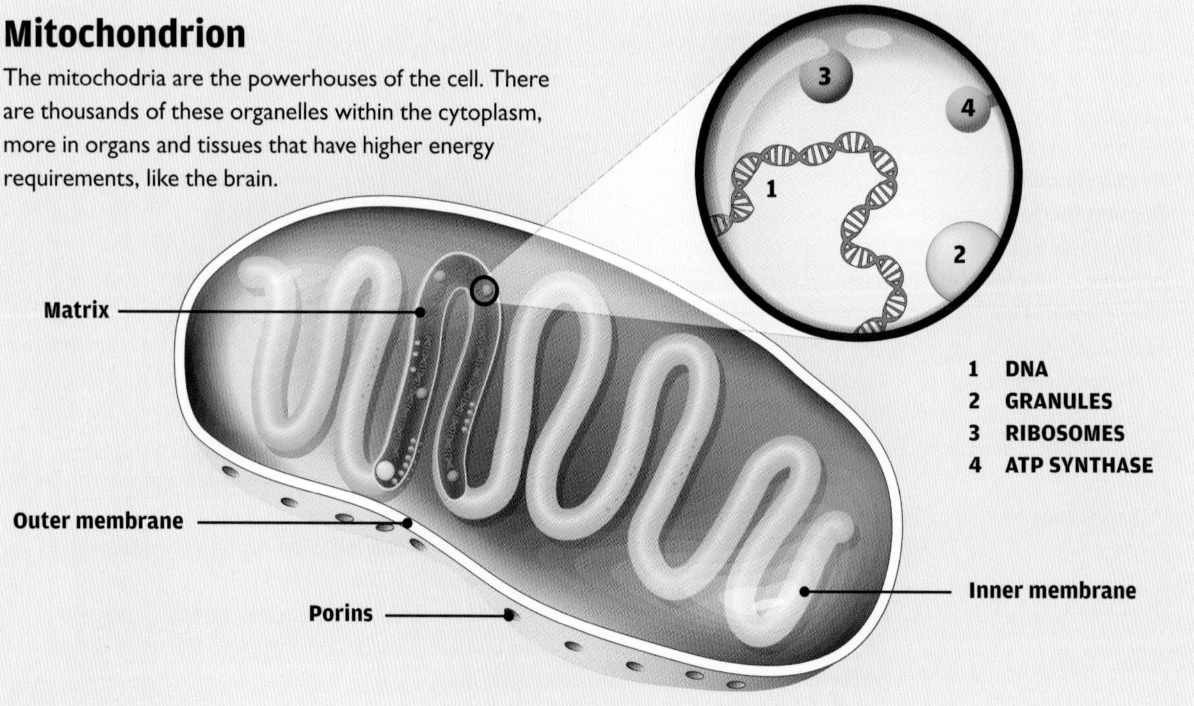

The level of heteroplasmy influences two major aspects of mitochondrial disorders: the likelihood of recurrence and the variability – both in the severity of a mitochondrial disorder and how it manifests. At a certain threshold level of heteroplasmy, around 60–80% with the balance in favour of the abnormally functioning mitochondria, a mitochondrial disorder will develop. The nature of that condition reflects the level of heteroplasmy in the tissues that have reached the threshold level. Because the phenomenon of heteroplasmy also applies to eggs, the ratio of normal to abnormal mtDNA in one egg can differ to the ratio in another, even though they came from the same woman. This means a mother can have children who both have a mitochondrial disorder, but with different manifestations and severity. Equally, one of her eggs can have a very low level of abnormal mtDNA, so the child this bears will not have a mitochondrial disorder; whereas another may have a high level, affecting that child much more significantly. Determining the ratio is extremely difficult and therefore so is predicting which tissues will be affected and to what extent, making genetic counselling very challenging.

Mitochondrial disorders

Because of unpredictable levels of heteroplasmy, disorders of mitochondrial function can be extremely variable, taking hold at any time, from the newborn to late in life. A mitochondrial disorder can be so severe because the tissues that rely upon high-energy production the most, the ones that are first to be impacted when mitochondria are not functioning efficiently, are the brain, the skeletal and heart muscles, and the liver. In very young children, mitochondrial disorders can cause blindness, delay in developmental milestones, muscle weakness and breathing difficulties.

Later on, different problems predominate, partly because heteroplasmy alters which tissues are involved and to what degree, and also because of different pathogenic variants in the mtDNA (different variants cause different diseases). Certain pathogenic variants happen recurrently, such as single base alterations or deletions and duplications of whole runs of DNA (in fact, just like we see in the nuclear genome) and this can lead to different medical conditions. Consistently characteristic health problems that raise the suspicion of a mitochondrial disorder outside of the newborn period are a variable combination of visual and hearing impairment, a disorder of movement and coordination called ataxia, shorter stature, and diabetes (both Type 1 and Type 2) occurring in later life. At the severe end of the spectrum, mitochondrial disorders are devastating and incurable. Often the load of mtDNA abnormalities in the eggs of an affected child's mother is so high that this condition will occur every time she has children. We can now use reproductive technology to circumvent the faulty mtDNA in the eggs, giving hope of having a healthy child, free from a severe mitochondrial disorder, to women in this situation.

Heteroplasmy

Mitochondrial DNA shows genetic variation, just like nuclear DNA. Pathogenic variants cause serious mitochondrial disorders. Because there are thousands of mitochondria in a cell, inclluding oocytes, only a few of them might have the variant – this is called heteroplasmy.

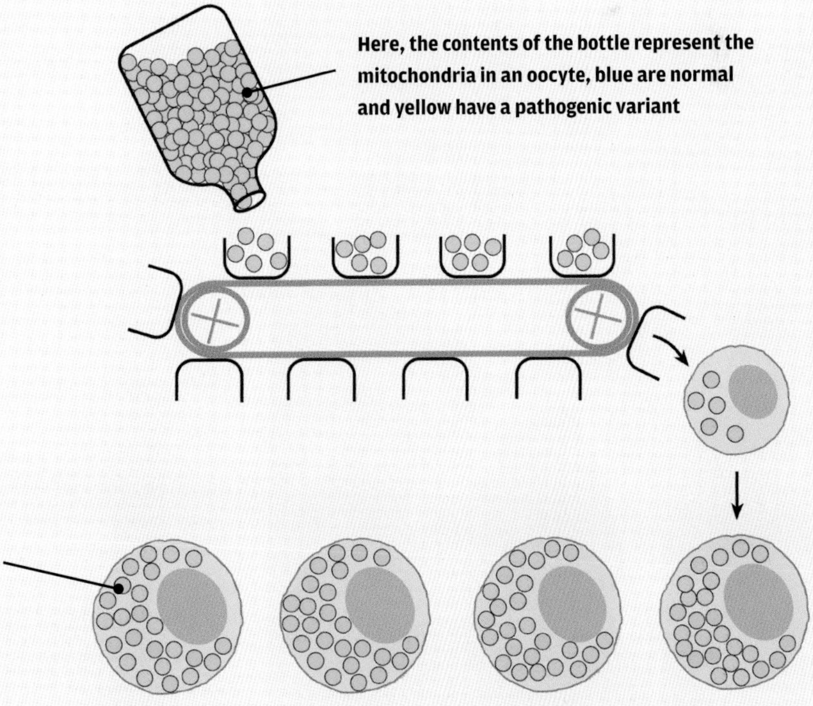

Skipping a generation

Sometimes a genetic condition seems to miss out a generation, but actually it will always have been there, written in the DNA. There are different explanations for why this apparent, temporary disappearance of a condition occurs.

Reduced penetrance and variable expression

Reduced penetrance occurs when someone has a disease-causing genetic variant, but they do not actually have the condition. In other words, having the variant does not make it 100% certain that a person will develop its symptoms. This occurs in some adult-onset conditions, such as susceptibility to certain cancers (breast, ovarian and prostate, among ot). Similarly, variable expression might have come in to play. Here, the genetic change can show itself physically in different ways, being particularly mild in its effect in one relative compared with another, such that it might not be obvious that it is there at all – in other words it *seems* to skip a generation.

This can happen in the brittle-bone disease condition, *osteogenesis imperfecta*: one person might break bones again and again throughout life, whereas a relative might just have problems hearing at one particular time in their life. Both have the same change in their DNA, but it has manifested in different ways.

▼ Alzheimer's disease, characterised by abnormal plaques forming on nerve cells, is a late-onset disorder to which some people might have a genetic susceptibility

Hidden in the maternal line

X-linked conditions, transmitted through a gene found on the X chromosome could be another explanation for an apparent generation skip. Here, the female relative linking one affected male to another, such as her father with her son, seems not to have the condition. She is, in fact, an 'obligate carrier', passing on the pathogenic variant she has inherited, but not necessarily expressing the condition herself.

The imprinting cover-up

Another, more unusual reason for a condition 'skipping' is the 'parent of origin' effect in genetic imprinting – involving those few hundred genes where we need both parents to contribute a gene copy. For example, the genetic condition may show itself only if the gene copy from the father is missing – and then only being apparent in his daughters, not his sons. His unaffected son, though, may go on to have daughters with the condition, as they did not inherit the key imprinted gene from him. It seemed to skip a generation, but the underlying genetic issue was there all along.

Recessive roulette

Autosomal recessive conditions may appear to skip generations when really genetic variants are still at play. In practice, this would happen only in conditions with a relatively high carrier frequency in the general population or in families where there is significant inbreeding or consanguinity. Here, one generation may have the recessive condition, such as cystic fibrosis or an inborn error of metabolism. All of their children will be carriers, such that it can be passed on to the next generation, but only if their partners are also carriers, missing out the generation in the middle.

Predisposition to late-onset disease

Whereas some genetic conditions start at birth or in childhood, others will show themselves only in adult life. These are so-called late-onset diseases. Generally speaking, these genetically predisposed conditions are neurological in nature or cause certain types of cancer. The late-onset neurological conditions are usually not treatable, can become progressively worse over time and may be life-limiting.

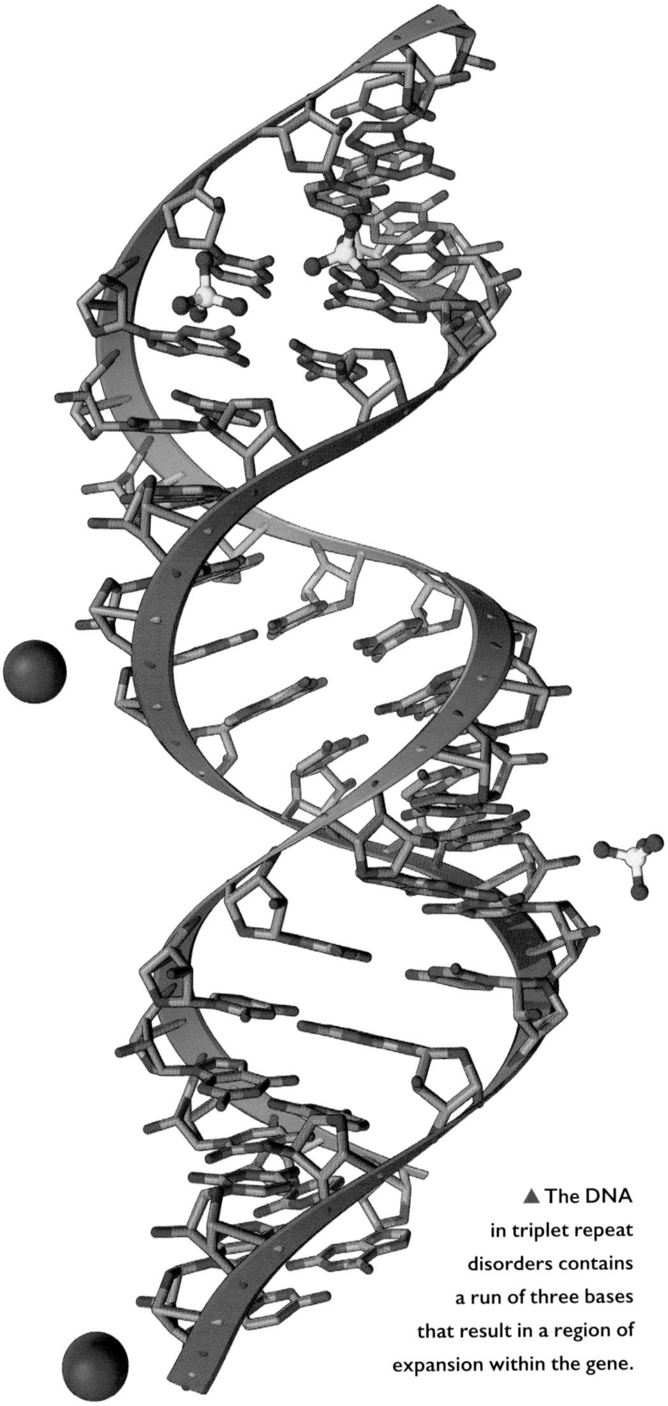

▲ The DNA in triplet repeat disorders contains a run of three bases that result in a region of expansion within the gene.

Huntington's disease

An example of a genetic disorder that falls into the category of being severe, late-onset and incurable is Huntington's disease (HD). It is a devastating neurological condition with an average age of onset at 40 years, a time when people have usually already had children. It causes three main problems: involuntary movements, memory loss and mental health problems (such as a personality disorder like schizophrenia). It is progressive, becoming steadily worse over the course of about 10 years. It is life-limiting and currently there is no cure. As the specific gene defect that causes this AD condition is known, it is possible to do predictive genetic testing of unaffected relatives with 50:50 risk.

Triplet repeats and genetic anticipation

A very specific kind of genetic variant, called a triplet repeat – a run of three nucleotide bases (CAG) repeated many times over – causes HD. The three bases form the genetic code instruction (codon) for the amino acid glutamine. The HD-causing triplet repeat occurs in the *HTT* gene that makes the protein huntingtin. We still don't fully understand what role huntingtin plays in the brain, but when a run of glutamine is added to it, its function is significantly altered, enough to cause HD.

The genetic test for HD counts the number of repeats. Everyone has a small number of repeats, usually below 36. But when the number reaches 40, HD is predicted to develop. (There is a 'grey area' between 36 and 39 repeats, when the predictive value is not certain, so it is difficult to say if and when HD will occur.) There is not an exact link between the number of repeats and age of onset, though, but generally speaking, the higher the number, the earlier it will happen. In fact, a childhood form exists where the number of repeats is around the 120 mark.

The number of repeats is different in us all because the genome can be unstable in certain regions, as we have already considered with DNA profiling. In HD, there is a specific kind of genomic instability called genetic anticipation, where the number of repeats tends to increase from one generation to the next, especially if it is passed down by the father.

Genetic counselling for predictive tests

Most genetic conditions that are classed as late-onset and severe are untreatable, which means that the medical profession encourages those who are thinking of undergoing predictive tests to also have a specific kind of genetic counselling that will prepare them for the outcome.

Genetic counselling asks questions such as: who you would tell if you get bad news? How might it affect your job and what are the prospects of getting life insurance? Equally, the impact of good news can be unsettling – for example, if you have built your life on the false assumption that you will at some point develop a life-limiting genetic condition (and so made decisions that seem frivolous or unworthy now that you have no reason to believe your life will be cut short); or if you receive good news but a close relative's news is bad (so-called 'survivor guilt').

Predictive tests

The first step in a test is to establish the precise pathogenic variant that occurs in the family in order to cause the disease – and this means testing someone in the family who already has it. Only then can the right person have the right test to answer with as much certainty as possible whether or not he or she will develop the condition later in life.

Every person who wants a test has a right to have one. However, it's important that we also respect the rights of those who don't want to know. Sometimes testing one family member means inadvertently testing another. Take, for example, a man who wants to know whether or not he will develop HD, knowing that his maternal grandfather had the condition, but not knowing whether his mother and then he had inherited it. If he has the test and it reveals bad news, it is inevitable that his mother has the condition too – and she may not want to know. For this reason, it's important that families talk to each other before undergoing any kind of predictive testing, and establish the parameters of their discoveries.

▼ We share our genes with our relatives, which is important to remember in genetic testing.

Woody Guthrie

The life of Woodrow Wilson 'Woody' Guthrie (1912–1967), the iconic US singer-songwriter, was plagued by family tragedy. Although those tragedies seemed coincidental, in retrospect are probably explained by the fact that Huntington's disease (HD) ran through his family. In his autobiography, *Bound For Glory,* Woody describes how his mother, Nora, would have uncontrollable mood swings and she would contort her face and body, both of which we now know are typical signs of HD. She was hospitalised when Woody was 14 with 'dementia and muscular degeneration'. When she died two years later, no one knew she had a genetic disorder that Woody and his siblings could inherit. In fact, Woody convinced himself that he would not have it because he mistakenly believed it could not be 'pass-on-able', as he put it, from a mother to her son.

Woody travelled around a great deal. He married three times and had eight children. He began showing signs of the condition himself, something that initially mystified him, but it eventually became clear that he too had his mother's 'dizzy disease'. Woody died of HD aged 54. Two of his daughters died of the disorder, too, both aged 41.

▲ **Woodrow Wilson 'Woody' Guthrie, December 1949, before the signs of HD became clear.**

Woody's ex-wife, Marjorie, set up the Huntington's Disease Society of America after his death. Today, it provides support for people with HD and their families, raises awareness of the condition and funds research into treatment.

▶ **Predictive genetic testing looks for a pathogenic variant within the DNA before it has any physical effects.**

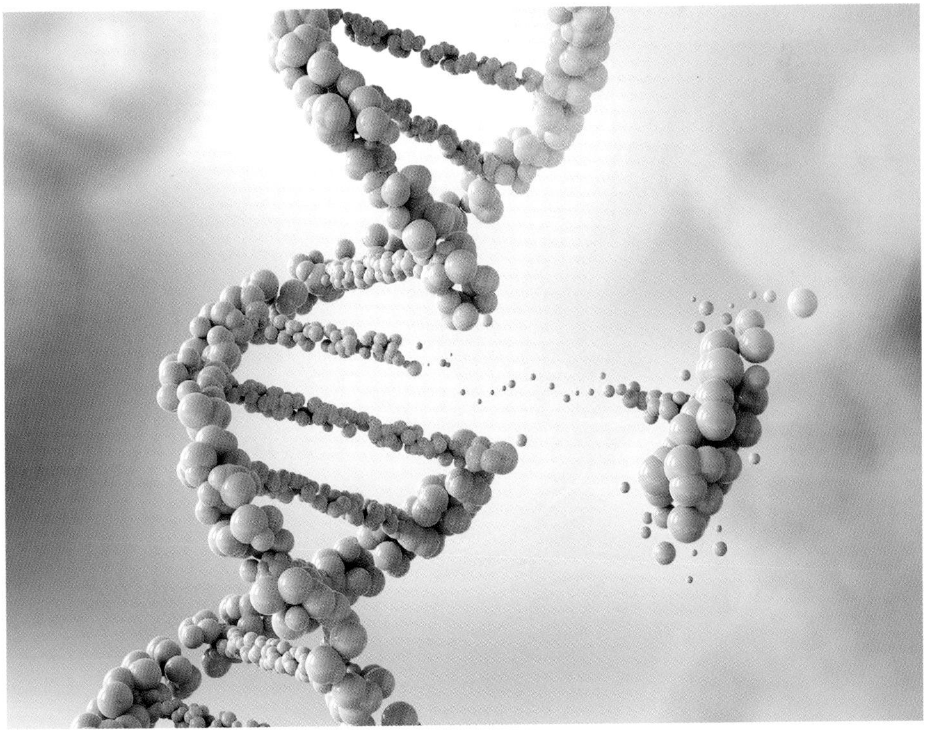

Rare diseases & common complex disorders

Asthma, allergies, diabetes, heart problems and migraines are all examples of complex diseases, meaning that the body needs an environmental trigger to expose the condition.

In conditions like this, rather than there being a causative change in one single gene, there are multiple, much more subtle variants, each of which – individually – has only a small, even negligible effect upon the cause, but in combination with other small effects triggers the disease. And there may be thousands of these susceptibility markers to look for, making genetic testing logistically very difficult and probably futile – considering the family history and estimating the chance of having the condition is often much more informative.

The genetics of autism

The prevalence of autism appears to be on the up. Worldwide figures suggest it was 1 in 2,000 the 1980s, when very little was known about it; in 2014 it was more like 1 in 68 (the UK prevalence is 1 in 100). So is autism becoming much more common and could DNA help us understand why?

Autism is actually a spectrum of issues affecting learning, behaviour and social communication. It varies considerably in how it affects different people. The most severe forms can be obvious from early childhood, but it is usually not formally diagnosed until the ages of between two and four, when young children develop more advanced communication skills.

Repetitive behaviour traits, such as rocking, hand-flapping and playing with toys in a specific, often unusual way, as well poor eye contact and being rigid with routines, are other characteristics of the autistic spectrum. It does not always impact adversely upon development, cognition and intellect, but when it does it can have a profound effect. Other people who are described as being 'on the spectrum' may even have an advanced intellect, often being labelled as having Asperger syndrome.

Multifactorial conditions

Autism is an example of a multifactorial condition, one that is a combination of both genetic and environmental causes. It is clear in some families where more than one person has typical traits that there must be a stronger genetic predisposition. And it is five times more common in boys than in girls – several genes and susceptibility regions on the X-chromosome have been identified to help explain this observation. In general, if one child in the family has autism, the chance of another also having it is 3–5% with a 10% chance of there being more 'spectrum-like' issues instead. But is it getting more common? Are there more environmental triggers today than there previously were?

Certainly autism in better recognised today and the extreme ends of the spectrum seem to be getting further apart, such that more children and even late-presenting adults are diagnosed as having autism. There are, though, no obvious environmental links to account for the increased prevalence. Any claimed links with the combined MMR (measles, mumps and rubella) vaccine are scientifically unfounded (and damaging – seeing an increased prevalence in serious measles conditions among children).

Pathogenic variants in a handful of single genes have been linked with autism, but these are rare associations. Markers in over 1,000 genes have been identified as having possible links too, most with only a tiny effect, although perhaps collectively resulting in an increased susceptibility. Many of the genes have a role in brain development, making them seem potentially good candidates when we're looking for a genetic explanation for autism spectrum disorders. However, that observation is, in reality, most likely explained by improved awareness and diagnosis of the condition.

Genetic testing and the law

One of the most important tenets of practising medicine is to uphold confidentiality at all times. Trust between a doctor and their patient is fundamental to well-being. But situations can arise that require striking a balance between the respect for a patient's right to confidentiality and the duty of care to protect them and others, when that confidentiality can lawfully be broken.

By the very nature of their cause, many genetic conditions have far-reaching consequences for a whole family, not just for one individual. Once the results of a genetic test highlight a high risk of a serious late-onset condition for the patient's close relatives as well, to whom does the doctor have a duty of care and can they break their patient's trust to inform those family members or must they maintain confidentiality no matter what?

Predictive testing for a late-onset genetic condition requires supportive genetic counselling that also takes into consideration the potential implications for family members. Doctors take care to minimise the potential to 'harm' others, avoiding, for example, a situation where genetic testing in one person might unwittingly disclose the medical status of another.

Rarely, this can break down and there are a handful of examples where a family has challenged a doctor's position of having ultimate duty of care only to their patient. Relatives who have made significant life decisions based on an assumption of inheriting a disease, or who could have taken precautionary steps to prevent the onset of a disease had they known, have provided the basis for legal proceedings in the past.

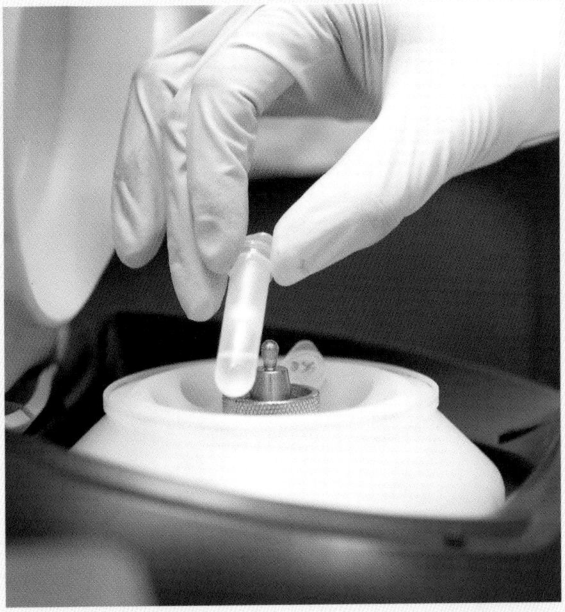

The law is clear when it comes to a doctor's duty to breach confidentiality, in the case of serious crime, serious communicable disease or their patient poses a serious risk to themselves or to others because they are unfit to work or drive. But in the situation where another family member is at risk of a genetic condition that their relative refuses to disclose to them after they receive their results, the legal expectations of healthcare professionals are less well-defined. Genetic testing presents a unique situation in this respect.

Genetics and insurance

In 2019, the Association of British Insurers (ABI) drew up the Code on Genetic Testing & Insurance, which it intends to review every three years to ensure it remains up to date with medical and ethical developments. The code states that when the life insurance sum is under £500,000, insurance companies must not pressurise any of their customers into taking genetic tests, predictive or diagnostic; they must not ask for or take into account a predictive genetic test result; and they must not take into account the result of a predictive test undertaken as part of scientific research. For applications above £500,000, those who have previously undergone genetic testing and have tested positive for Huntington's disease must reveal the result and are, at that insurance level, not covered by the code.

Genetic predisposition to cancer

Cancer occurs when cells grow abnormally and out of control. Cell growth is under the control of a vast array of genetic signals, depending on their type and location in the body. Starting with fertilisation of the egg at conception, cells grow and divide, die and are replaced. There is constant turnover, with new cells growing from older ones. And every time this happens, the cell has to replicate its entire genome perfectly – if this is not as accurate as it can possibly be, mistakes will start to appear in the DNA. These errors in the genetic code can damage the proteins they produce to override the normal processes that control cell growth and result in cancer.

There are checkpoints built into our genetic code to ensure that the entire set of DNA is replicated faithfully every time a cell regenerates. These checkpoints detect and correct any mistakes. In fact, we have special proteins designed to repair DNA that has been damaged by environmental factors, such as ultraviolet radiation and tobacco smoke (exposure to which can cause skin and lung cancer, respectively). Sometimes, though, these quality-control processes go wrong or they cannot combat the environmental damage effectively.

There are two sets of genes that we can directly link to cancer formation when irreparable damage occurs. The first set is made of up of suppressor genes, which should offer protection; and the second is made up of oncogenes, which drive cell growth.

Tumour suppressor genes

These genes regulate growth to protect cells from replicating out of control. *BRCA1* (Breast cancer 1 gene) is an example of a tumour suppressor gene. Everyone, males and females, has two copies of it because one copy is inherited from each parent. The role of the *BRCA1* protein is to repair damaged DNA. If there are pathogenic variants in *BRCA1*, it cannot carry out its repairs, meaning that damage goes unchecked and uncorrected and cells can grow out of control. Tissues that are particularly prone to cancer when *BRCA1* is altered are tissues of the breast and ovaries in women, and the prostate gland in men (although this is much less common).

Environmental damage

Exposure to elements in our environment that can damage DNA brings a higher risk of developing cancer. Our cells have a built-in safety system to protect the DNA, and repair any alterations to the DNA sequence. But it can be overcome, especially after repeated exposure to a damaging agent like cigarette smoke.

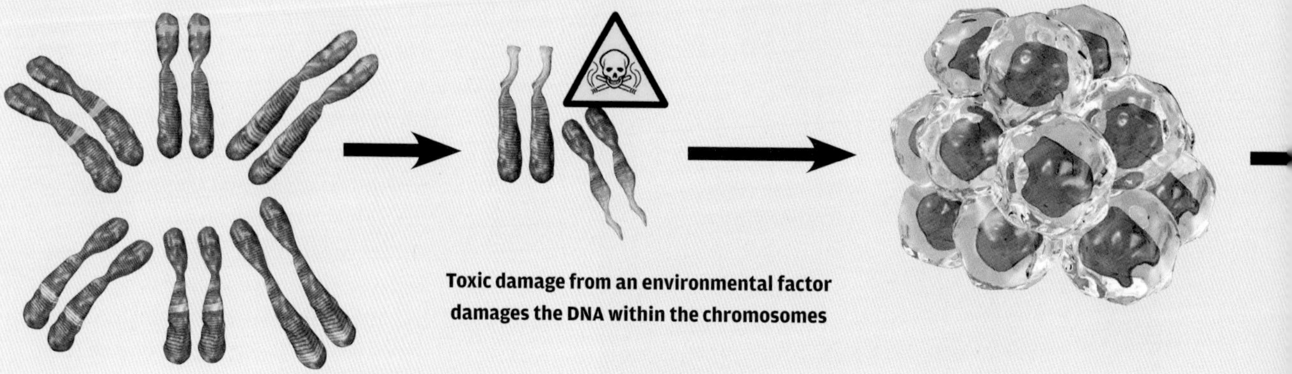

Toxic damage from an environmental factor damages the DNA within the chromosomes

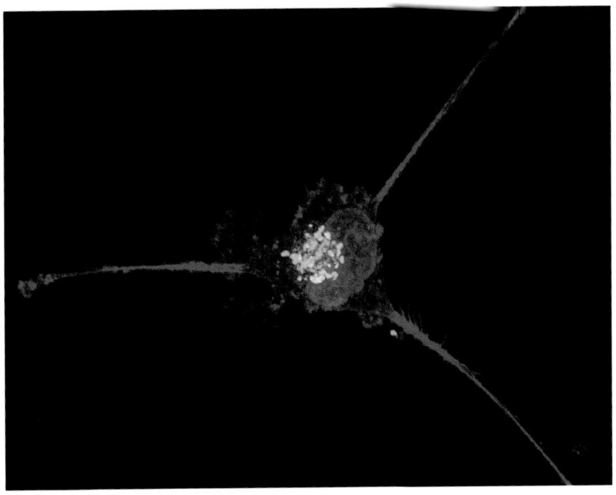

▲ **If genes designed to protect us from cancerous changes are altered, cells can grow out of control.**

Oncogenes

Proto-oncogenes form a group of genes responsible for controlling normal cell growth and division. If changes occur within a proto-oncogene, its effect upon the cells change and it can no longer work efficiently to control cell growth. The newly formed oncogene drives the cells to divide uncontrollably, causing cancer. For example, *RET* is a proto-oncogene that produces a protein responsible for controlling the signals the body needs for the normal growth of nerve cells. In particular, it influences the growth of the specialist nerve cells that are found in the intestines. These specialist cells make up the enteric nervous system and the autonomic nervous system. When a pathogenic change alters *RET* so that it becomes an oncogene with cancer-forming potential, a group of conditions called multiple endocrine neoplasia results. This group includes rare cancers affecting the glands, such as the thyroid and adrenal glands.

Inheriting a cancer gene

Problematic genetic changes in genes are rare. The statistic that one in two people will have cancer in their lifetime is more the result of cancer caused by exposure to damaging environmental factors, such as tobacco smoke and an unhealthy diet, than because of genetic changes. However, there are some telltale signs in a family's medical history that can hint at a genetic predisposition to cancer.

If there are more people who have had cancer in your family than you would expect by chance; if there are certain cancers that seem to cluster together in the family (like breast and ovarian cancer); if family members develop cancer at a much younger age than usual (below 40, for example); and in the case of breast cancer, if someone in your family has had cancer in both breasts and if breast or ovarian cancer has affected very close members of your family (such as your mother or your sister), you may be more genetically susceptible. Clinics set up to assess your individual risk of cancer based on your family history can help you decide whether you and your relatives should have testing.

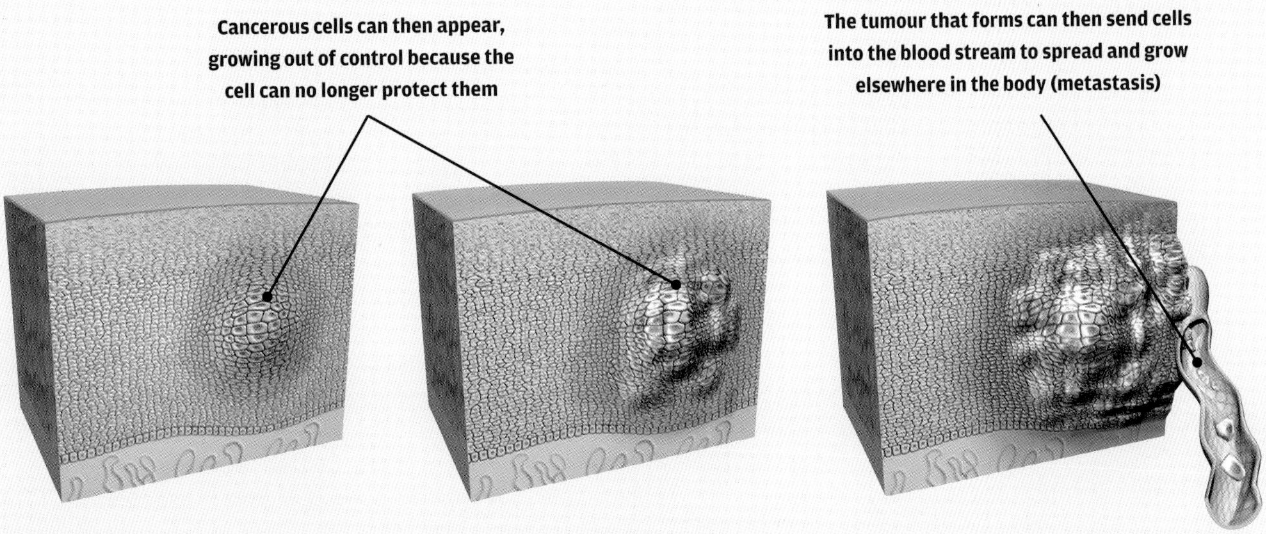

BRCA1

Most cancer-susceptibility syndromes, like those linked with *BRCA1*, for example, are inherited in an autosomal dominant way with a 50:50 chance of transmission from one generation to the next. But not everyone who has a damaging change will develop breast and ovarian cancer, although the chances are significantly higher when compared to women in the general population.

By following the health of women until the age of 80 it has been estimated that 72% of women with a *BRCA1* pathogenic variant will develop breast cancer – this compares to 12% (one in eight women) in the general population by this age. The risk of ovarian cancer is 44% if you have a pathogenic variant in *BRCA1*, compared with 1.3% in the general population. If a man carries a pathogenic variant in *BRCA1*, there is a 1% risk he could develop breast cancer.

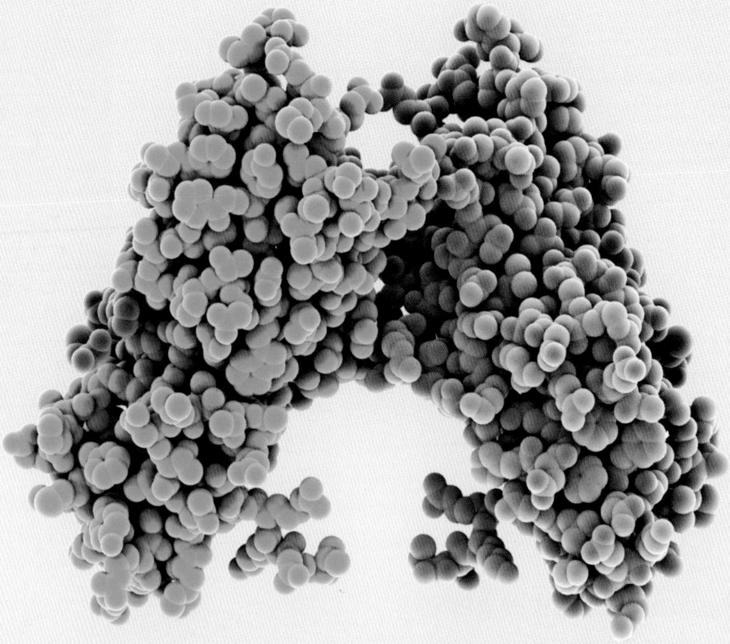

▲ Computer model showing the human breast and ovarian cancer susceptibility protein *BRCA1* protein complex (green, blue) complexed with an *ATRIP* peptide (pink).

Tumour markers

The presence of a set of tumour markers in biopsy samples provide another reason to think that there could be an underlying genetic susceptibility to cancer. An important step in diagnosing breast cancer, for example, is to look for progesterone and oestrogen hormone receptors, and for human epidermal growth factor receptor 2 (HER2), key information for planning cancer treatment. HER2 is a protein that controls the normal growth of breast cells. If the body makes too much of this protein, the cells grow out of control, driving breast cancer development. Drugs, like Herceptin, block the receptors. Hormone-receptor blockers can also be effective, but only if the tumour is receptor positive. If they are triple negative and none of these receptors are present on the tumour cells these drugs will not work. Triple negative tumours are more common when there is a genetic susceptibility to breast cancer, for example associated with pathogenic variants in *BRCA1*.

Predicting cancer

When the nature of a family's medical history raises the suspicion of a genetic predisposition to cancer, testing someone who has had cancer can identify the cause behind the susceptibility. We can test for a panel of genes, like *BRCA1* and related genes (*BRCA2* and *PALB2*), say, depending upon the type of cancers that have occurred.

Other cancer-susceptibility genes are linked to some types of bowel and womb cancers (Lynch syndrome), prostate cancer in males and breast and ovarian in females (*BRCA2*), and there are a handful of very rare cancer syndromes, such as von Hippel-Lindau (*VHL*), and Li-Fraumeni syndrome (*TP53*), where a number of exceedingly rare cancer types can occur. The occurrence of unusual patterns of cancer in a family might raise the suspicion of a rare genetic defect as the cause.

Modifying the risk of cancer susceptibility

Once we have identified that a person has an increased risk of cancer, as a result of genetic effects, we can take steps to reduce cancer risk. For example, if a woman has a genetic predisposition to breast cancer, she will know to examine her breasts regularly. Furthermore, she can begin breast screening earlier than it's offered to the general public (mammograms and MRI scans will usually begin at around the age of 40, rather than 50). Sometimes consultants will recommend surgery to remove the breast and ovaries, along with the fallopian tubes, which can significantly reduce the risk of developing breast and ovarian cancer for women at very high genetic risk.

The Angelina Jolie Effect

In 2013, Angelina Jolie, the Hollywood actor and humanitarian, wrote in *The New York Times* about her decision to undergo removal of both her breasts by double mastectomy. Aged 37, Jolie opted to have DNA testing to look for the *BRCA1* gene variant that had caused her mother and maternal aunt to develop breast and ovarian cancer; her mother had died aged 56. She was told that she had inherited the damaging *BRCA1* variant and with it a significantly increased risk of developing these two cancers. Jolie therefore made the choice to undergo major risk-reducing treatment because she felt the stakes were simply too high to leave to chance.

Publicity around her genetic test and consequent choice to undergo risk-reducing surgery triggered the 'Angelina Jolie effect', a boom in the number of women seeking medical advice for possible breast-cancer susceptibility genetic testing. It did not, however, extend to increased demand for similar risk-reducing surgery, which is sometimes combined with removal of the ovaries and tubes, too.

▼ Laptop screen showing results of a home DNA test (23andMe). Because of the wide-reaching medical implications, *BRCA* genetic testing requires expert genetic counselling support.

Genetics and reproductive medicine

A growing number of reproductive technologies and different DNA tests are making it increasingly easy to screen for a likelihood of passing on a genetic condition to an unborn child. At what point during pregnancy these tests can happen depends upon the condition doctors are looking for.

Prenatal diagnosis

Prenatal diagnosis is a way to find out during a pregnancy whether a baby has a genetic disorder. Usually, doctors will undertake prenatal diagnosis when medical history or a routine test indicates that the baby has a high chance of developing a certain condition. Doctors will take a sample of the baby's DNA and perform a genetic test, such as chromosome or individual gene analysis.

Chorionic villous sampling (CVS), amniocentesis, foetal blood sampling (FBS) and collecting free foetal DNA (ffDNA, which circulates in the mother's blood stream) are all ways in which doctors can obtain the unborn baby's DNA. The first three involve invasive procedures, using ultrasound to guide the long needles into the mother's womb. Sampling ffDNA, though, is non-invasive – it is simply taking a blood sample from the mother. It is known as non-invasive prenatal testing (NIPT).

Prenatal diagnosis

Prenatal diagnosis requires genetic testing of DNA from the foetus. It can be obtained from different sources during a pregnancy. Invasive tests, like amniocentesis, require inserting a needle into the womb. It must be carefully positioned to withdraw some of the amniotic fluid. This is done under ultrasound scan guidance.

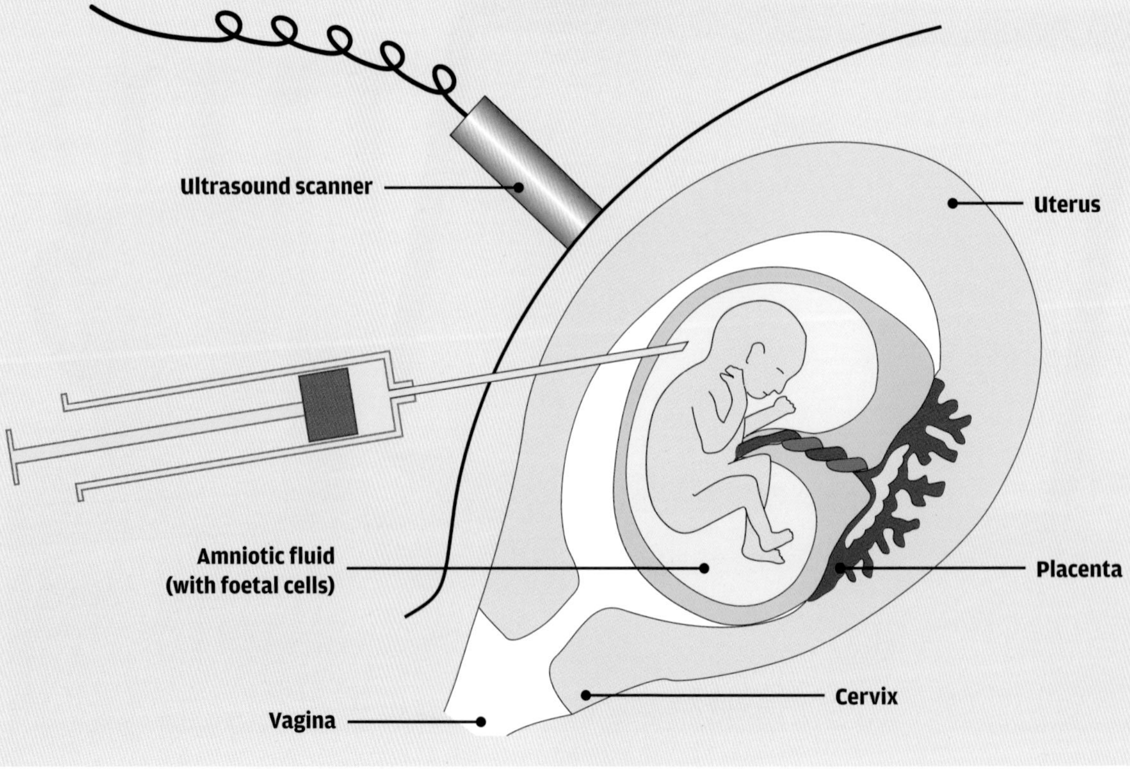

Is it a girl or a boy?

NIPT enables us to detect the sex of the baby long before a woman can have an ultrasound to tell her whether she is having a girl or a boy. Detecting Y-chromosome DNA in the mother's blood points to a male foetus. The practical use for this is that only babies identified as male by ffDNA analysis need then to have invasive testing for X-linked disorders.

ffDNA is also useful in diagnosing autosomal dominant conditions that could either be *de novo* in the foetus or inherited from the father. It is more difficult to use this test when the mother has the dominant condition, however, since extracting the baby's DNA from her blood sample will also isolate her own DNA, revealing her dominant gene defect and overwhelming the much lower concentration of foetal DNA.

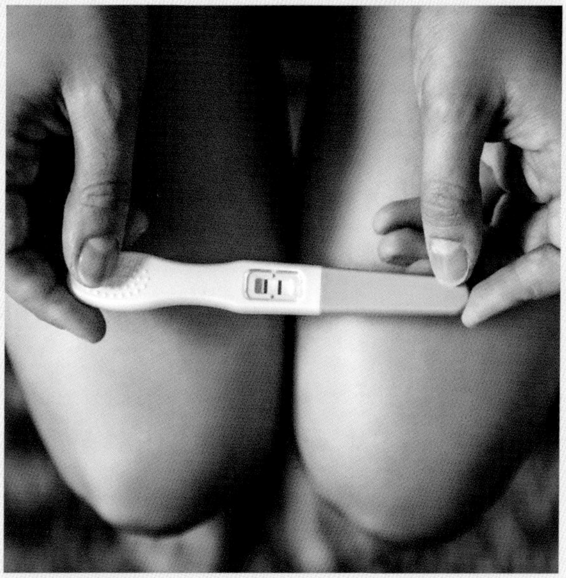

Invasive versus non-invasive foetal DNA sampling

The chorion is a membrane that forms in the embryo, the collection of cells that result after the fertilised egg divides and starts to generate more cells. It goes on to form the placenta, providing the growing baby with nutrients and oxygen, and removes waste products and carbon dioxide. Because the chorion, along with its stalk-like villi that project into the cavity of the womb, is embryonic in origin, it contains DNA from the baby. The CVS procedure, performed between 11 and 14 weeks of pregnancy, removes a sample of the chorion and its villi.

Withdrawing a sample of amniotic fluid, the liquid that surrounds a developing foetus, from inside the womb is called amniocentesis (usually performed around 15 weeks of pregnancy). The fluid contains cells from the baby, shed from areas such as the urinary and gastrointestinal tracts and from the skin. The cells contain the baby's DNA.

FBS (performed at around 18 weeks of pregnancy) involves taking a blood sample, containing the baby's DNA, directly from the baby through the umbilical cord.

NIPT has two main advantages over all these other ways of getting foetal DNA. First, it is non-invasive, which reduces the risks to the pregnancy as a result of having a procedure. Second, a woman can have it any time from very early on in pregnancy until the end, long after other procedures might trigger early labour and delivery.

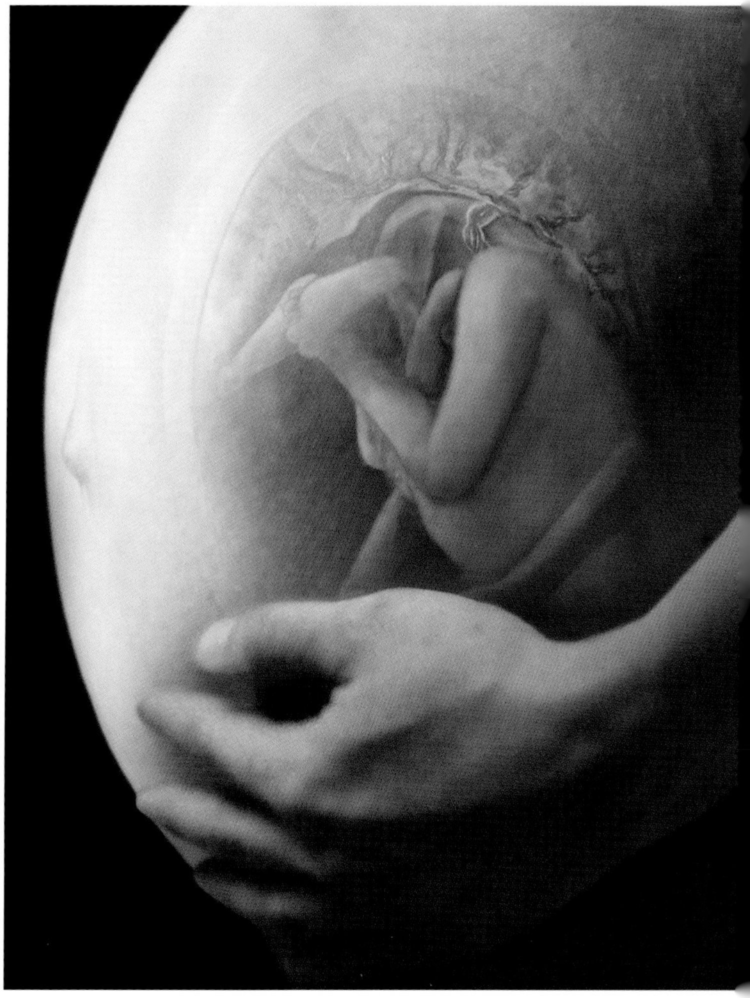

▶ Non-invasive prenatal testing is not associated with a risk of miscarriage.

Routine pregnancy screening

All pregnant women are offered ultrasound scanning (USS) at around 12 weeks and at 20 weeks of pregnancy (although in some parts of the UK, the 12-week scan may not be available on the NHS).

The combined test

The 12-week scan measures the thickness of a layer of fluid at the back of the baby's neck called the nuchal translucency (NT). A fold that is thicker than about 3.5mm can indicate that the baby is at risk of a chromosomal anomaly, like trisomy 21 (Down syndrome). (The thicker the fold, the greater the risk of the anomaly.) Around the same time, doctors will perform a blood test on the mother's blood that examines a hormone called free beta-hCG and the protein PAPP-A, both of which the woman's body produces during pregnancy.

The results of the hormone and protein tests and the result of the nuchal translucency test make up the combined test, which gives a much more accurate estimate of whether or not there is an extra copy of a chromosome that could cause abnormality than the NT test alone. If the risk comes out as high, doctors carry out rapid analysis of the chromosome number, using CVS, amniocentesis or NIPT. The analysis looks in particular for more than the normal number of two copies of chromosomes 13, 18 or 21 and X/Y. Any other chromosomal trisomies at this stage in pregnancy are not compatible with life – which means that the pregnancy will certainly miscarry very early on, so testing for them is not necessary.

▼ Tests routinely offered in pregnancy include an ultrasound scan to look for markers of a chromosome disorder.

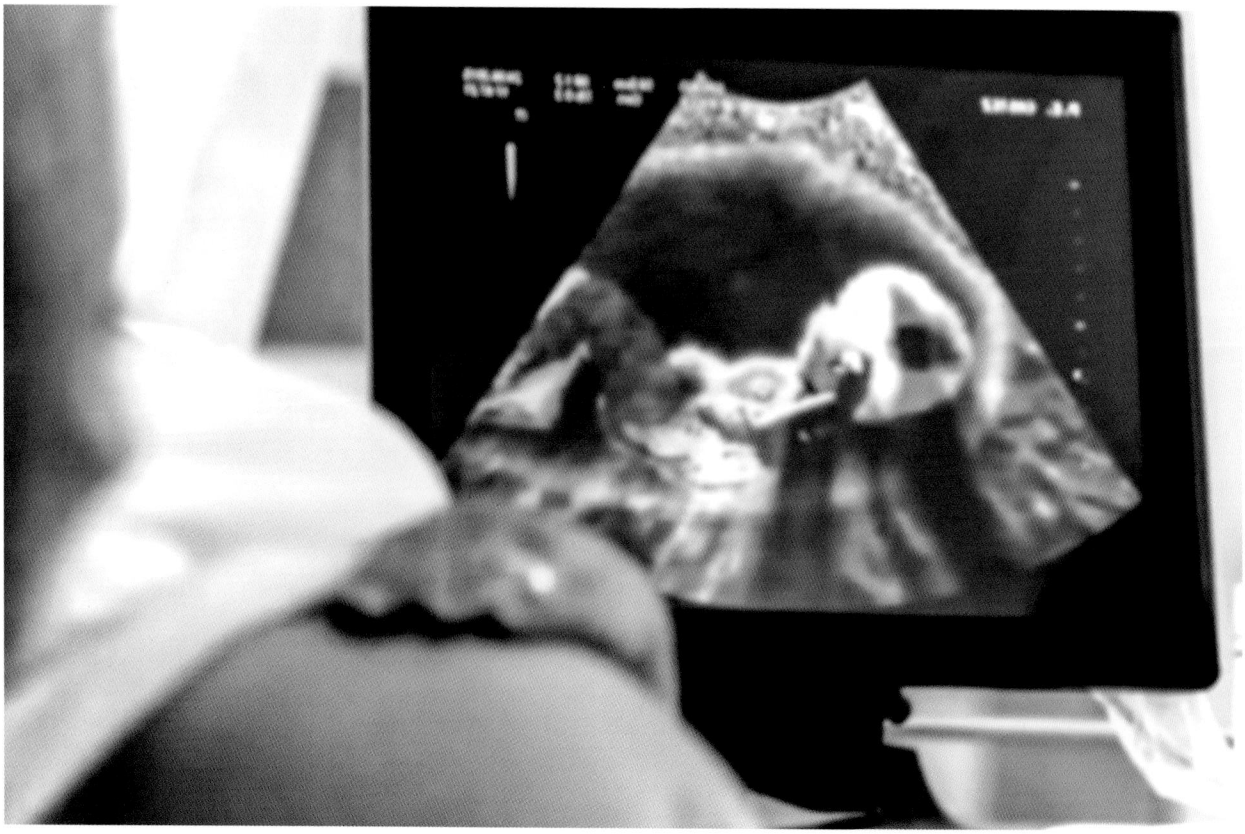

Down syndrome karyotype

Rather than the usual number of 46 chromosomes, there are 47 in Down syndrome. There is an additional chromosome 21 – instead of two copies, there are three. This is why it is also called Trisomy 21.

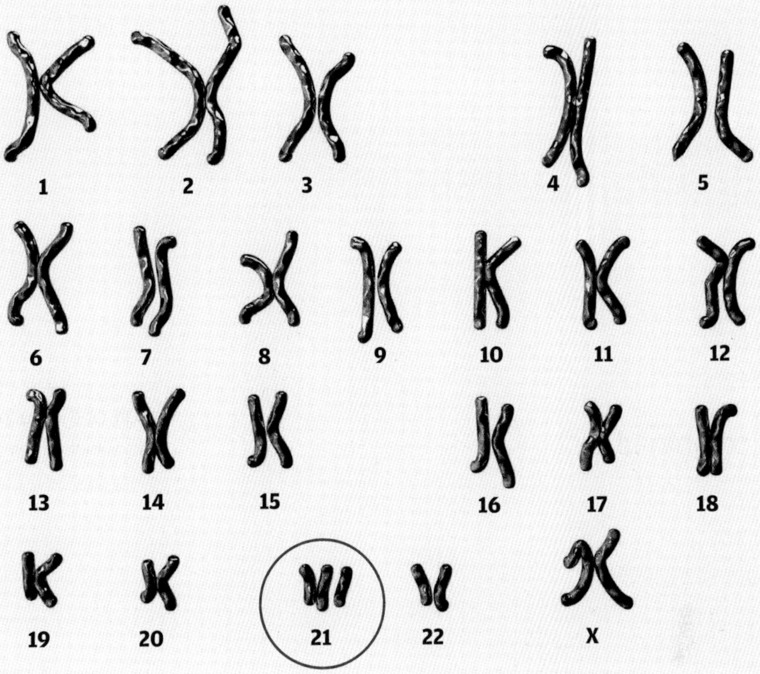

The anomaly ultrasound scan

At 20 weeks of pregnancy a woman will have an anomaly ultrasound scan. Many expectant parents see this as the opportunity to find out if they are having a boy or a girl, but it is actually intended to identify congenital anomalies and problems with the baby's growth. In the event that the sonographer can see multiple congenital abnormalities, such as a heart defect or short bones in the arms and legs, or one kidney instead of two, the family may receive chromosomal analysis on their baby. Other signs include an absent bone in the bridge of the nose and small pockets of fluid in the brain called choroid plexus cysts. Although it is possible to use NIPT for these tests, it is more likely that doctors will perform an amniocentesis, which looks for abnormalities in more than just chromosomes 13, 18 or 21, as any chromosomes may have abnormality at this stage.

Normal results

If the results of prenatal testing do not show up an issue with the chromosomes that would explain the defects a sonographer can see on a scan, or the results are not conclusive, doctors will consider other possible explanations, such as a syndrome or a single gene defect. They will use a panel of genetic tests to test single or multiple genes and attempt to determine which suspected syndrome is the cause.

Sometimes, though, it's not possible to get answers. When results remain unclear, medical experts will carefully monitor the pregnancy and check the well-being of the baby, either for reassurance that nothing is getting worse or to detect any further worrying changes.

Raised parental age – the mother

Chromosomal aneuploidy, having a different number of chromosomes from the usual 46, is more common in older pregnant women. The risk of aneuploidy increases with advancing maternal age: by the time a woman is 40 her risk is 1 in 85, increasing further to 1 in 35 at 45 years old.

Why should this be the case? The process of separating each of the chromosome pairs to form the egg starts way back during a baby girl's foetal development, and it is not completed until just before that girl's body (having reached puberty) releases the egg during a menstrual cycle. So the older you are, the longer your eggs have been hanging around waiting to complete that final step of reduction division. And the longer they wait around, the more likely it is that something will go wrong with them.

Prenatal genetic counselling

All parents who discover that their baby may have a congenital abnormality are offered genetic counselling to help them come to terms with the diagnosis and decide whether or not to terminate the pregnancy.

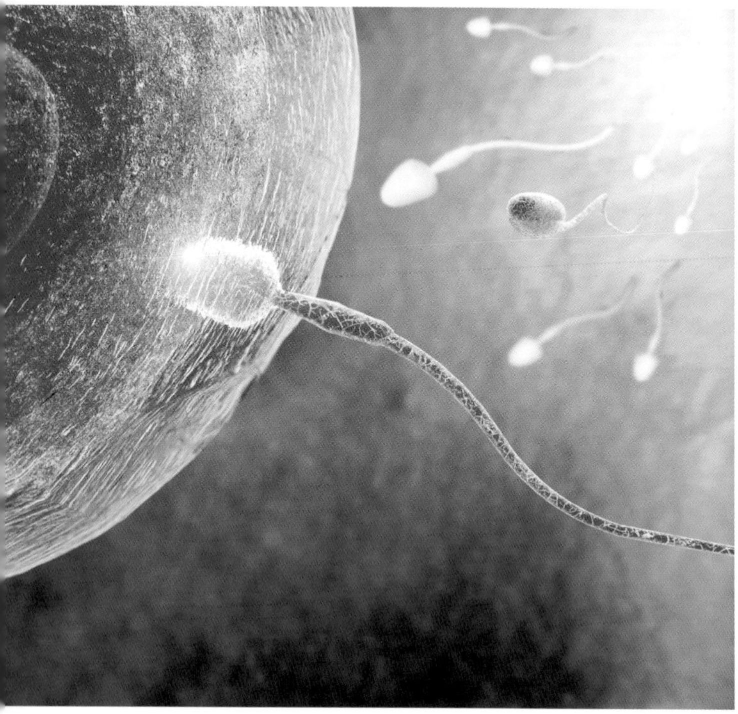

Chromosome 21 is the most commonly involved chromosome in trisomy, but it can also happen with chromosomes 13 (Patau syndrome) and 18 (Edwards' syndrome). Routine pregnancy screening specifically looks for the possibility of these conditions, because they are associated with major congenital abnormalities, like severe heart and brain defects.

Raised parental age – the father

For fathers, new dominant genetic disorders are the more likely consequence of advancing age. Achondroplasia is one example. It is the most common form of dwarfism with a genetic cause, although it only occurs in 1 in every 20,000 people. It is the result of a very specific genetic variant in the fibroblast growth factor receptor type 3 (*FGFR3*) gene.

For 80% of people with achondroplasia, there is no previous family history – it started in them for the first time, a *de novo* genetic change that then becomes a dominant trait that they can pass on the next generation. So how did they get it? And why does this particular change seem to happen again and again, even when no one else in the family has it?

It seems some genetic variants bring a selective advantage to cells in the male germline, the ones that produce sperm. Studies of the testes in men of different ages show in older men there is a much higher concentration of gene variants, ones that when passed on to a child result in one of a handful of new dominant conditions typically seen in the offspring of older fathers. There is something about this particular set of variants that means the cells are much more competitive in the sperm-producing stakes, which is why this is sometimes known as 'selfish selection'.

Routine screening tests, such as ultrasounds, in pregnancy can detect some of these conditions in the babies of older fathers, but unlike aneuploidy screening for older mothers, there is no specific testing.

Advantages of prenatal testing

The advantage of prenatal testing over prenatal diagnosis is that once the woman is pregnant, the prospect for a successful pregnancy is much greater than using prenatal diagnosis and waiting to see what happens. The downside, though, is that assisted conception does not always guarantee a pregnancy that leads to a live birth – success rates vary depending on a number of factors, including the age of the mother. As a technique, prenatal testing brings great hope and much expectation, but it has major limitations, too.

Preimplantation genetic diagnosis

For some people, waiting until a pregnancy has become established and then having genetic tests is not acceptable, especially when faced with uncertainty. Preimplantation genetic diagnosis (PGD) offers greater control for parents whose offspring are at risk of a genetic disorder.

PGD is a reproductive technology that provides genetic testing of embryos created using assisted conception technology, such as *in-vitro* fertilisation. The idea is to create embryos outside of the woman's body, allow them to divide to at least the eight-cell stage and then genetically test two of them at a time (others may be frozen while tests are carried out) and implant only embryos (usually up to two at a time) predicted not to have the condition. Of course, there is a risk that no embryos will be suitable for transfer.

Duchenne muscular dystrophy, cystic fibrosis or a complex chromosome defect are all reasons why doctors may recommend PGD testing – it is reserved for use with serious genetic disorders only. The Human Fertilisation and Embryo Authority (HFEA), an independent regulator appointed by the UK government, makes sure clinics abide by a code of practice to ensure the safety of patients going through these treatments, prevents unacceptable use of reproductive technology and runs a licensing system for PGD.

Mitochondrial transfer

A mitochondrial disorder is a disease that prevents the mitochondria (the energy powerhouses of a cell) from functioning properly, in turn preventing the cells from having enough energy to do their jobs properly. It is a lifelong, but relatively rare, genetic condition that can affect any organ of the body, depending upon which cells are affected.

A procedure used to avoid passing on a serious mitochondrial disorder, mitochondrial transfer uses a specially adapted method of reproductive technology. The technique was developed to prevent transmission of the mother's mitochondria to her offspring, while ensuring she can pass on her nuclear DNA, the main bulk of her genetic code. The approach requires a donor egg with healthy mitochondria, its nucleus removed and replaced with the mother's nucleus; and sperm from the father, which are used to create an embryo for transfer to the mother. The procedure has led to the term 'three-parent baby' and has transformed the prospects of having healthy children in families previously devastated by serious mitochondrial disorders.

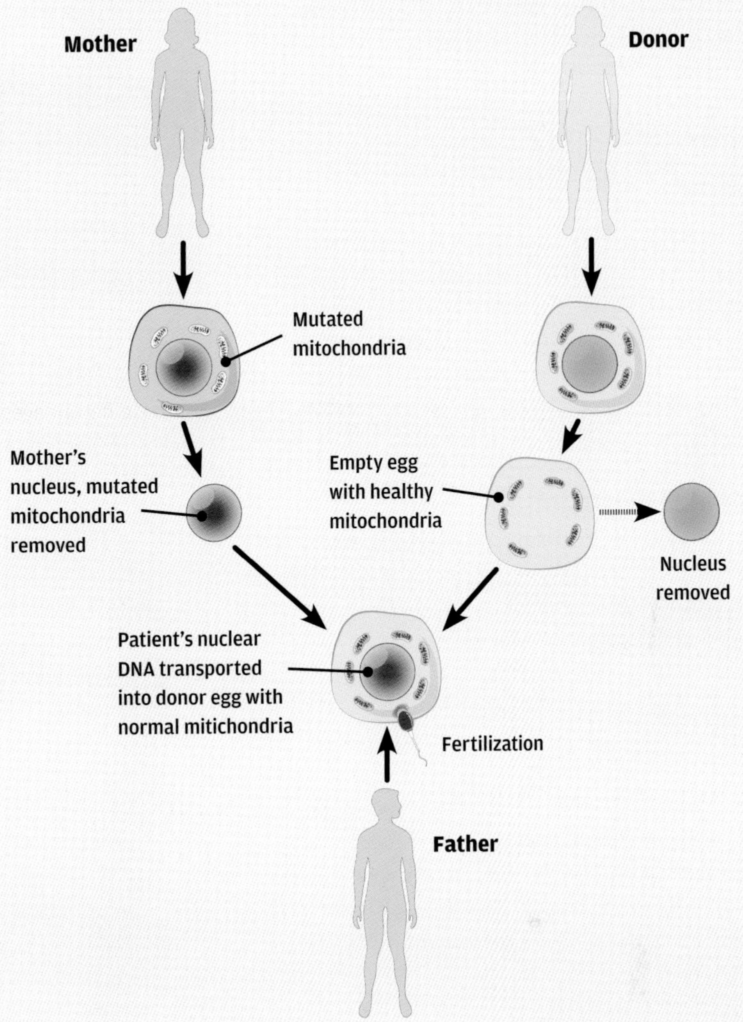

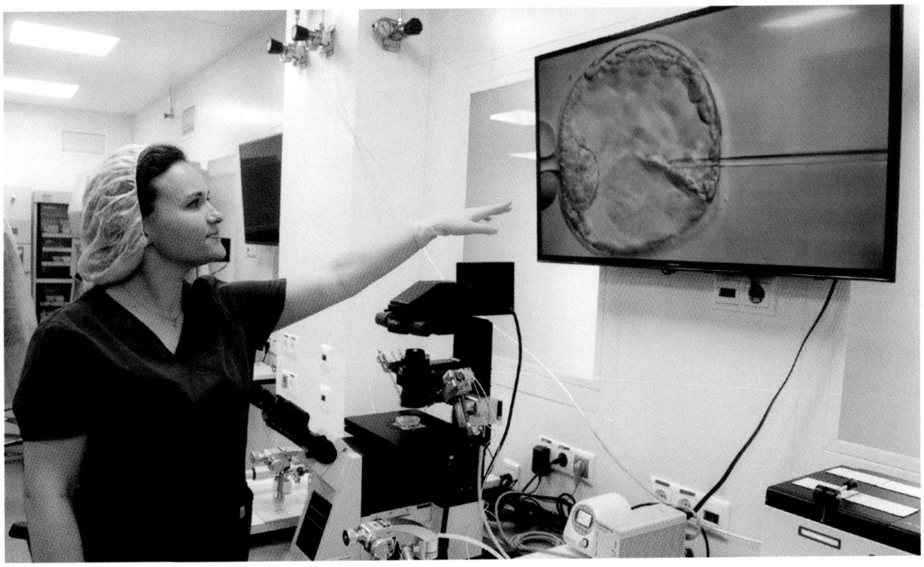

▼ Embryologists require sophisticated technology to biopsy such a tiny embryo.

Preimplantation Genetic Screening

Preimplantation Genetic Screening (PGS) uses the same technology as PGD, but to a slightly different end. The aim is to check embryos for chromosome abnormalities, a major cause of miscarriage in a pregnancy and a reason why IVF treatment might fail. The intention is that screening out embryos with chromosomal aneuploidy from the outset will improve the chance of a successful pregnancy. As this is more of an issue for older mothers, some clinics offer PGS to women aged over 37. There is no conclusive scientific medical evidence yet to say that PGS does improve the prospects for older women, but it might increase the success rate for women under 37 who do not have any prior history suggestive of a chromosome problem.

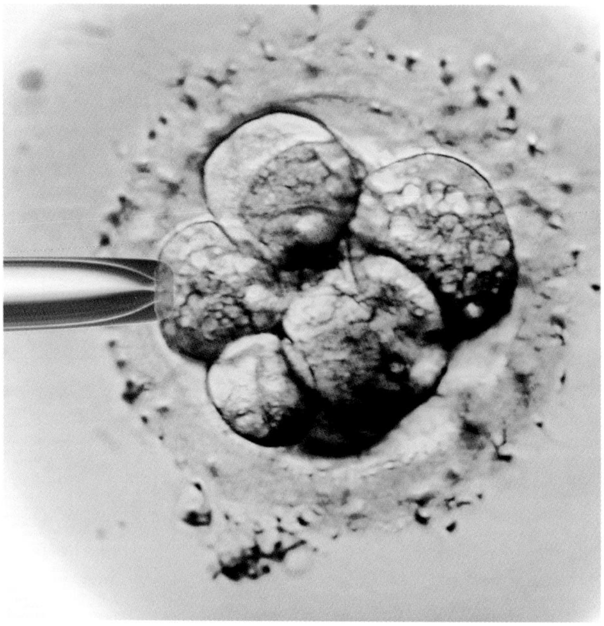

▲ **The microscopic embryo is made up of a small number of cells, one or two of which are taken for genetic testing in PGD.**

Prenatal exclusion

Prenatal exclusion is a method of genetic testing to find out if a foetus has a high-risk gene copy of a condition like Huntington's disease without revealing if the parent at risk has inherited it from their affected parent. For example, if a pregnant woman whose father has HD wants to make sure

Prenatal exclusion testing

Prenatal exclusion testing is a method to prevent passing on a genetic condition without revealing to the 'at-risk' parent if they have inherited the condition themselves.

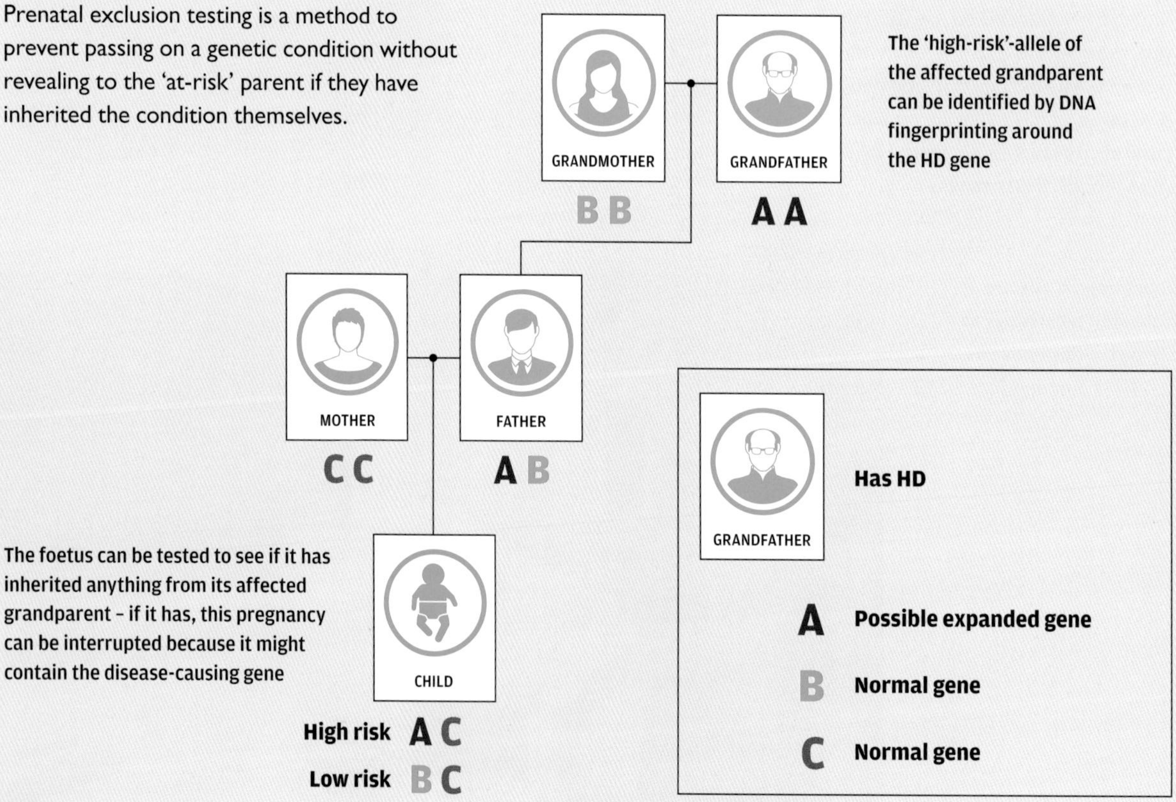

Genetic testing in children

In the eyes of the law, young people are deemed to have the mental capacity required to make important decisions about themselves, such as consenting to medical procedures and tests, from the age of 16 years. Until that time, their parents or legal guardians are appointed to represent them and their best interests. This applies to genetic testing as well. Children might be in the position to have a genetic test for one of two reasons: to diagnose a genetic disorder or to find out something about their future. The two situations are ethically very different.

We can use generic analysis as a diagnostic medical test to see if a genetic condition explains concerns about a child's health or development and helps guide their medical care and therapy needs. (Testing to make a diagnosis of a genetic condition does not have to be DNA-based – it can be a biochemical test on blood or urine, an X-ray or a tissue biopsy looking for pathological signs that indicate the presence of a genetic condition.)

When it comes to testing for a late-onset condition, especially one for which there is no treatment, or finding out if a child is a carrier for a genetic disease, there are other factors to consider. Testing removes a child's autonomy to make decisions about their own future. We have already come across the complexities that result from predictive testing for HD, even when adults go through it. Research studies demonstrate that there is negative impact and stigmatisation for children who undergo carrier testing, when the results in fact have no bearing upon that child's health. There is no doubt that testing has a place in children's medicine – but only when we carry it out with a child's best interests at heart and provide the correct support before and afterwards.

Adoption and genetic testing

Before children are adopted, they generally go through a number of assessments to check on their overall health and cognitive development. This is to identify their potential future educational and healthcare needs and to ensure an appropriately matched family is found. Increasingly, local authorities are under pressure to seek genetic testing as part of these assessments, sometimes under the misapprehension that the information from genetic testing is always definitive in respect of the child's future.

However, stigmatising and commodifying children on the basis of their genetic code is fraught with difficulty and paediatricians need to weigh up the benefits of genetic testing (for example, to the prospective adoptive parents) with the harm it can do.

she does not pass on the disease, but does not want to know if she has inherited it, she can have this type of prenatal test.

The prenatal test relies on DNA fingerprinting to identify and rule out passing on either of the gene copies that originated from the affected grandparent by excluding them. Researchers analyse the DNA fingerprint around both the woman's father's HD gene copies and compare it with the DNA of the foetus. Importantly, the test is designed just to identify which of his daughter's gene copies came from him. DNA from the foetus (usually taken by CVS or amniocentesis) reveals if either of the gene copies originally came from the woman's father. If one of them was passed on, the woman would be offered a termination.

Of course, there's risk – the procedure may upset the pregnancy of a healthy baby, which can lead to miscarriage.

Special considerations in genetic testing

Ethical, legal and social issues frequently surround genetic testing. Privacy and informed consent are cornerstones of the care provided by professionals who unlock the information held within the code. And there are certain situations when special consideration should be given before conducting a genetic test.

CHAPTER 7

What happens next?

We've come a long way since Mendel first described inherited traits in his sweet peas, Watson and Crick discovered DNA (with a little help from their friends) and we finally cracked the genetic code. Genetics is no longer a secret science – not a day goes by without some reference to it in the news or online. It seems to be becoming increasingly relevant to our daily lives and it is easier now than ever to access our own genomes through direct to consumer tests and as part of routine medical care. So what lies over the horizon? As we take the genomics revolution forwards and harness its power to full effect and maximum output for the greater good, we must also be aware of potential misuses and meet these head-on in order to prevent future genetic catastrophe.

Precision medicine

A relatively new field, precision medicine is set to transform healthcare – although for now we don't have the infrastructure or intricate knowledge to put it in fully in place. The aim is to take into account genetic variation and its interplay with lifestyle and environmental factors in order to prevent disease and improve treatments. It will replace the 'one-size fits all' approach to medical care, focusing instead on the individual.

Pharmacogenomics

Pharmacogenomics is a specific branch of precision medicine that holds the key to explaining why some people respond better than others to a particular drug treatment, some do not react at all and others might even have adverse drug reactions and serious side effects. It combines pharmacology (the study of drugs and how they work) and the science of genomics. The basis of these individual differences in response lies in the variation within the genetic code, especially in genes that make proteins responsible for drug metabolism.

The metabolic processes that alters the chemical structure of a drug, either to turn it into the active form required for treatment or to clear it from the body, are generally performed by enzymes. Subtle differences in the genes that make them can change the way they work, making them more or less efficient. Genetic variation could determine how easily a person's body absorbs the drug, how cell receptors pick it up and react to it, how quickly the body can process it to become active and how long it is in this state before the liver and kidneys remove it. Understanding how all of these stages can be different from one person to the next because of differences in the genetic code means prescribing medications can become much more tailored to the individual.

▼ Pharmacogenomics is a branch of precision medicine that tells us how genes affect our response to different drugs.

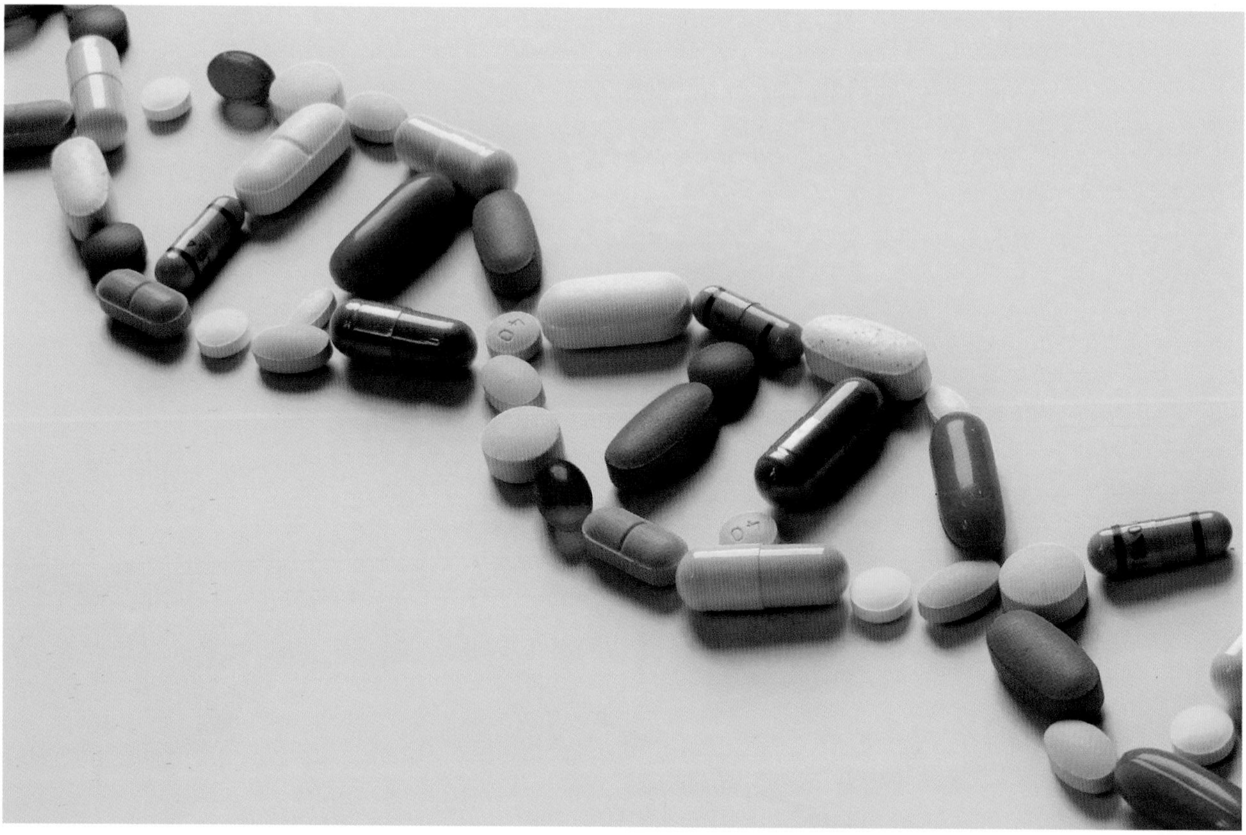

Personalised prescribing

A number of frequently prescribed drugs are known to have different effects in different people. Conventionally, doctors will prescribe a standard dose to everyone and wait to see if the patient responds (positively or adversely) to the medication. Pain medicines and antiepilepsy drugs, antidepressants, anticoagulants and immunosuppressant drugs are all medications treated in this way.

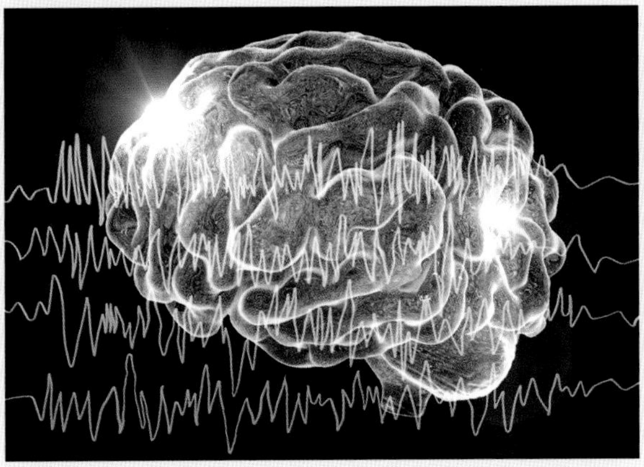

Pain medicines

This group of drugs includes codeine, tramadol, morphine and methadone, and we all respond differently to them, depending upon our unique ability to metabolise them – which is down to variants in our genes. In the future, personalised prescribing will make it possible to tailor prescriptions to a pain medicine that will be most effective and efficacious for our personal genetic make-up, at a dose that does not cause serious side effects.

Antiepileptics

Some antiepileptic drugs, such as carbamazepine, can cause unpredictable, severe adverse reactions, such as kidney and liver failure, in some people, but not in others. This effect is much more common in people from certain ethnic backgrounds, such as the Han Chinese, which suggests that genetics are at play. In fact, there is a particular variant in the genetic codes of the Han Chinese that we can use as an indicator of who will develop the side effects, so that doctors can tell, patient by patient, who should and should not be prescribed the drugs.

Adverse drug reactions and side effects

If you have ever read the small print on an information leaflet that comes with any medication you will have encountered a list of potential side effects as long as your arm. Some of them seem mild, like dry mouth, whereas others sound much more serious, such as a severe allergic reaction called anaphylaxis, which, without rapid treatment, can lead to death. These are known as adverse drug reactions. If you have then gone on to take that medication, you might have experienced side effects or you might not have. A very small number of people will experience the more serious reaction.

Most adverse drug reactions are unpredictable. So much so that one classification system once categorised them as 'bizarre'. But by collecting information about the way different people react to medications, good and bad, from thousands, if not millions, of patients and combining that with data from their genomes we can start to predict individual drug response more precisely. At the point of prescribing a drug, in the future, we will be able to look at the patient's genomic markers and identify which medication is most appropriate for their profile, which one will work the best with least side effects, and even which dosage to use. Researchers are already developing 'bedside' devices to do this – sometime not too far in the future, these will bring genetic technology directly to a doctors' surgery near you.

▼ Why some people have side effects to drugs and others don't could be written in our genes.

Gene therapy

A form of medical treatment for rare genetic disorders, gene therapy repairs faulty genes. For example, in the bleeding disorder haemophilia A, the gene that makes the protein 'factor VIII' essential to the body's clotting cascade pathway, does not work properly. If a patient suffering from haemophilia A undergoes gene therapy, he or she can receive a working copy of the appropriate gene in order for their cells to start making factor VIII. Equally, a person could receive gene therapy in order to inactivate a gene that is working abnormally, or even to introduce a gene that can fight disease.

Gene therapy in action

In order to replace a missing or a faulty gene, laboratory experts create a new working copy of that gene to deliver to the patient's cells. The new gene requires a kind of vehicle, called a vector, to get to where it needs to go. For this, gene replacement therapy uses viruses, the tiny microbes best known for giving us coughs and colds, and childhood infections like measles. The therapy capitalises on a virus's ability to invade (infect) the cells of a host and get them to make the proteins it needs in order to make us unwell. In gene therapy, though, the virus is stripped of the apparatus it needs to cause illness and is left just with the DNA it needs to invade the cells. Once inside the cells, the new DNA copy switches on and starts making the protein.

Gene replacement uses many different viruses as vectors, which specialists inject through the patient's vein into their blood stream. Some, like retroviruses, then have the ability to insert their DNA into the host's genetic code, and the replacement gene along with it. However, retroviruses may deliver their DNA anywhere in the body, which could damage the host's other genes. Other viruses, such as adenoviruses, invade the cell, but keep their DNA separate from the host's. Because adenoviruses can invade many different types of cell, they are useful for delivering gene replacement for a number of rare diseases.

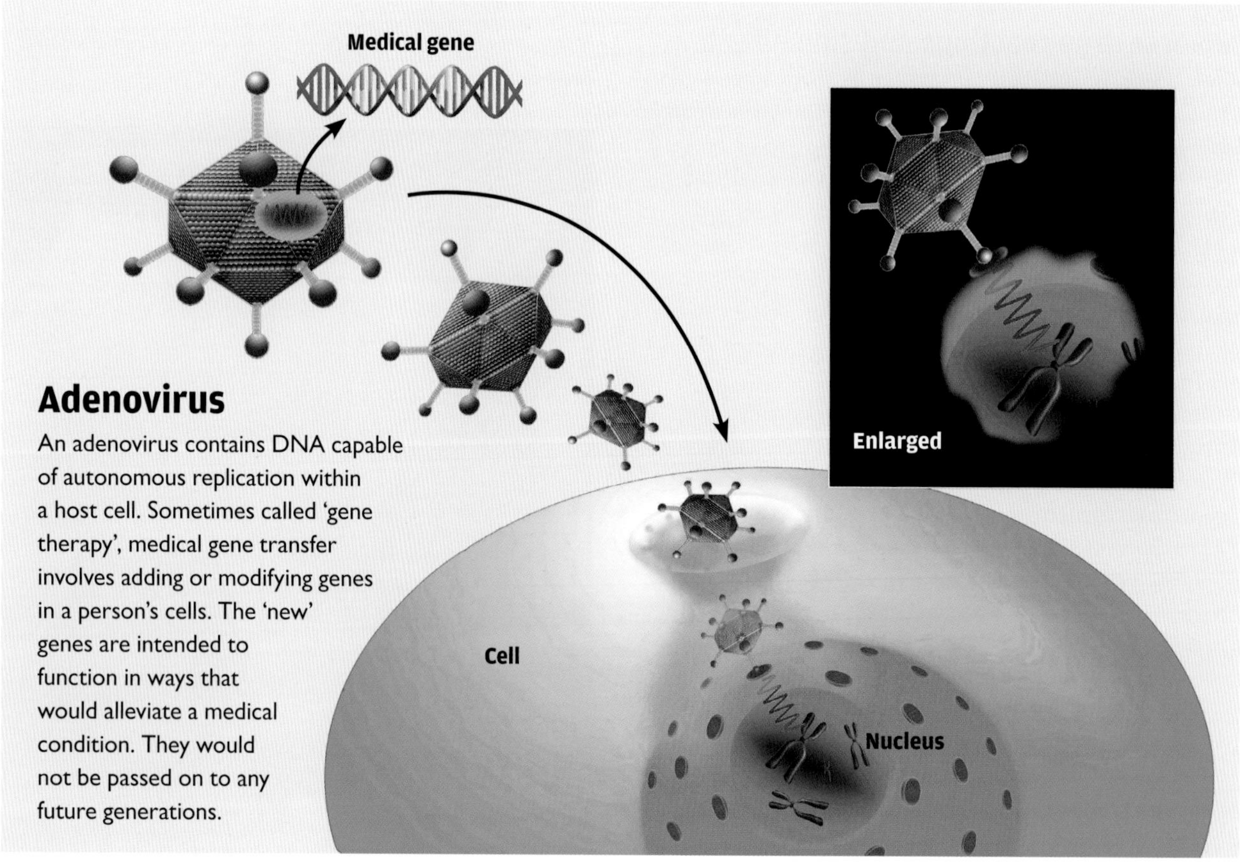

Adenovirus

An adenovirus contains DNA capable of autonomous replication within a host cell. Sometimes called 'gene therapy', medical gene transfer involves adding or modifying genes in a person's cells. The 'new' genes are intended to function in ways that would alleviate a medical condition. They would not be passed on to any future generations.

What can we use gene therapy for?

Conditions where the target tissue is easiest to access, such as those affecting the retina at the back of the eye, or where the gene fault occurs in a localised, easily accessible part of the body, are the most amenable to gene therapy. In patients suffering from retinopathy (genetic, degenerative damage to the retina), gene therapy may be able to restore some sight in some individuals – or at least as far as early research suggests. In cystic fibrosis, gene therapy delivers the replacement gene directly to the respiratory tract by inhalation. And gene replacement therapies for a severe neurological disorder that affects babies, called spinal muscular atrophy (SMA), are soon to become accepted treatments for this progressive and life-limiting autosomal recessive condition that affects one in every 6,000–10,000 live births.

What is the future for gene therapy?

The hopes for gene replacement therapy in the future are to treat the root cause of genetic disorders, namely to overcome faulty genes, and to find a delivery system that is both safe and effective. So far, this treatment has been developed for use in only a small number of rare genetic diseases, but the prospects for gene therapy are extremely promising and exciting – it could potentially treat multiple genetic disorders on a much larger scale.

Potentially, there are other uses for gene therapy, too. For example, we might be able to use it to kick-start new cell formation to encourage damaged tissues to regenerate. This could transform recovery following heart attack, for example, or help to repair damage to the spinal cord.

The ethics of gene therapy

The ability to introduce a new gene into human cells brings with it significant ethical issues. Concerns such as what constitutes good genes and bad genes, using the technology to create a race of genetically 'elite' individuals and the high financial cost of gene therapy are among the ethical issues this branch of medicine faces. As with other reproductive technologies, pre-emptive discussions with groups representative of society will be important in addressing questions like this and to make policies on a global scale, regulating its use for the benefit of everyone.

▶ The future success of gene therapy rests on finding safe and effective delivery systems. One possibility is the use of medical nanoparticles like these.

DNA cloning

DNA cloning is used to copy, or clone, a gene by inserting it into a circular piece of DNA called a plasmid. Viruses are then used to deliver working copies of the gene to the patient's cells, where the protein can be made.

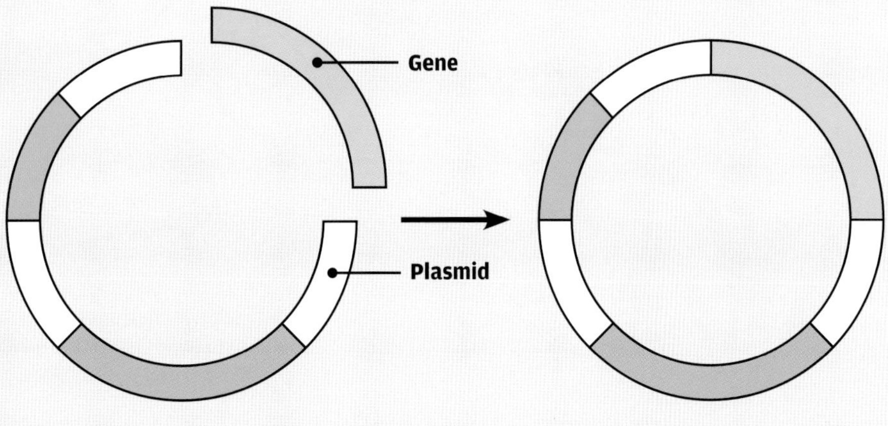

The epigenetics revelation

Genetics is the study of heredity, traits transmitted through the DNA from one generation to the next. However, what matters in epigenetics are the chemical tags interspersed throughout the genetic code that can impact upon the way genes work. In other words, epigenetics is the study of the chemical modifications to the DNA and the effects they have on the genome. Unlike with DNA itself, a small number of these tags are passed on to the next generation, but mostly the epigenetic slate is wiped clean going from one generation to the next, or so it is thought. The importance of epigenetics in understanding our health and well-being is becoming clearer and more significant every year.

Epigenetic modification of the genome

Epigenetics adjusts the workings of the genetic code without actually changing its basic blueprint. That happens by the addition of tiny chemical compounds to the DNA. There are two main ways these changes exert their effects: one is by changing the formation of the DNA strand and the other by switching on or off certain genes in certain cells.

DNA methylation

In DNA methylation, a methyl group of one carbon atom and three hydrogen atoms (chemical formula CH_3) attaches to a few strategically placed cysteine (C) bases in the genetic code. This has the effect of switching off or silencing genes, controlling when and where proteins are made. For example,

DNA methylation

DNA methylation occurs when a chemical called methyl (CH_3) is added to the strand of DNA. This lets genes be silenced or activated without changing the DNA sequence.

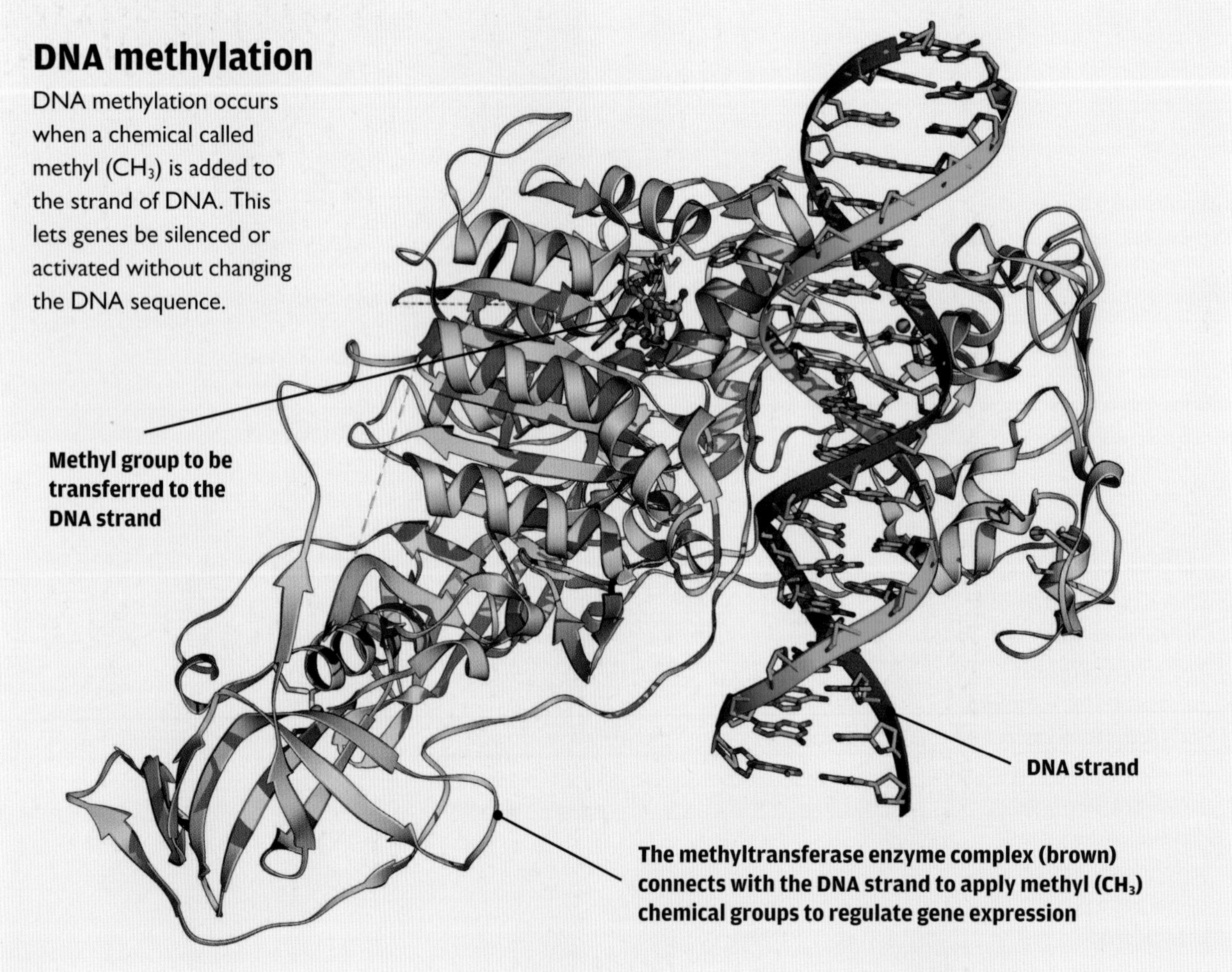

the haemoglobin-producing genes should be active in the bone marrow, but not in the brain, and foetal haemoglobin should be made by a foetus, but not generally after birth. Methylation is also the mechanism by which females inactivate one of their X-chromosomes – it is covered with methyl tags and silenced.

Methylation acting in this way controls the genes, but it also seems to have an additional function because our genomes become more methylated over time. That suggests methylation responds to environmental exposures, too. Exactly how, though, and just what information is hidden in the library of DNA chemical modification remains to be seen. We hope that the new science of methylomics will tell us more so we can start to understand the interplay between nature and nurture.

Histone modification

Histones are the tiny proteins that DNA wraps itself around to coil the double helix strands in sequentially tighter and tighter turns to form nucleosomes, chromatin and finally chromosomes. It also requires chemical modifications, a dynamic process involving chemical compounds such as acetyl (chemical formula CH_3), made up of a carbonyl group of one carbon atom and oxygen (CO) added to a CH_3 methyl group. A dynamic balance of adding and taking away acetyl tags from the histone proteins changes the way the DNA is folded, exposing or hiding the genes within and so altering which ones are active and which ones are not.

Epigenetics and ageing

Research has shown that our epigenetic profile changes as we get older. There are clear differences between the methylation pattern of an embryo, an infant, a young adult and an octogenarian. The concept of the epigenetic clock illustrates this change over time. It compares chronological age with biological age as determined by a few hundred methylation hotspots throughout the genetic code.

The premise is that over time, the epigenetic markers of environmental exposures accumulate and the degree to which our genome is methylated at certain genomic positions is a reflection of the extent to which we have been exposed to a particular factor – more exposure equals more methylation and therefore accelerated biological ageing. Why these effects should occur is still not clear, nor is exactly what the tests are measuring. Until we understand better the mechanism that links the methylation profile with the exposure, we will not know how to stop the clock, or at least slow it down, an especially attractive future prospect for those hoping to beat the relentless ageing process.

Unraveling DNA

If we unravel DNA from the chromosomes, we can see how it is packaged up and coiled around histone proteins. DNA methylation is also shown to see how different epigenetic markers work together.

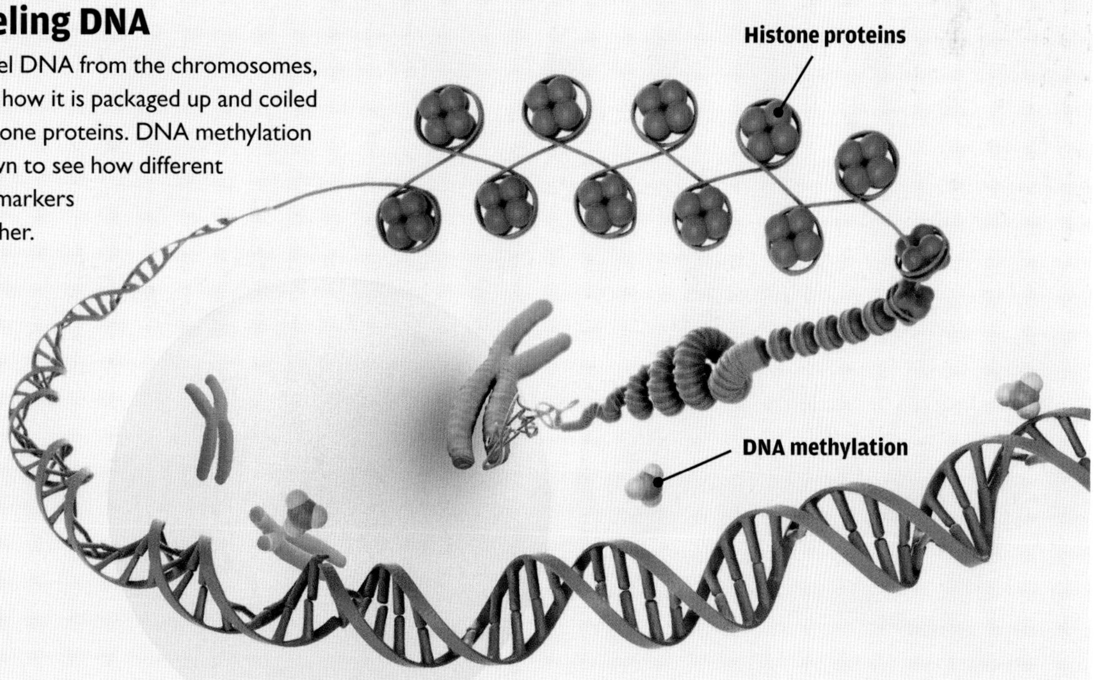

Why does epigenetics matter?

It is well known that actually identical twins are not fundamentally alike, yet they have the same genetic code. So what accounts for the differences? It has long been thought that the differences are related to the twins' environment – exposures to certain factors, such as exercise, sleep, illnesses and trauma (all of which we know influence epigenetics) are disparate between twins.

So in the context of nature versus nurture, epigenetics is emerging as the main driver behind the effect of our environmental exposures on our health. But it plays an important role much earlier on in life during embryonic development, too. It is a key factor in organ development and in determining what type of cells will form from the stem cells that have the potential to turn into a number of different kinds. And when things go wrong, the impact is significant – there a number of neurodevelopmental and growth syndromes that are the consequences of abnormalities in the way the epigenetic markers are laid down. Some of them may be reversible, paving the way for potential therapies to treat this very specific group of rare diseases.

Epigenetic markers, the chemical modification tags on the DNA, also seem to be linked with more common diseases, such as Alzheimer's disease, and some types of cancer, including breast, and mental health disorders.

There is gathering evidence that epigenetics might also have an association with autism and obesity. And the pattern of markers changes with time and with exposure to different environmental factors, such as cigarette smoke. Ageing affects our overall epigenetic profile because that too changes over time. The epigenetic clock is the perfect illustration of this.

Inter-generational transmission

Scientists hope that epigenetics will give up the answers to questions such as how does environmental exposure to certain factors like smoking and diet alter the fundamental workings of the DNA and what happens to these processes over the life course of a human being? But will the story stop there or might we also find there is a knock-on effect of our exposures for the next generation and maybe even the one after that?

The idea of inter-generational, also called trans-generational, transmission of traits is concerned with passing on physical characteristics not through the variants in the genetic code, but as the result of parental exposure to certain environmental factors that affect our own epigenetic profiles. The observation by David Barker in 1990 that poor nutrition early on in life, a reflection of the mother's nutritional status, during embryonic and early foetal development, can influence the subsequent health of that individual in adult life forms the basis of the Barker hypothesis.

◀ Identical twins might have identical DNA, but they are not truly identical – could epigenetics explain this?

The Dutch famine of 1944-1945

During the Dutch famine of 1944–45 in the German-occupied Netherlands, 20,000 people died of starvation. It was a brief yet devastating period, ended by the liberating in May 1945 of the Dutch people at the end of the Second World War. A number of women were pregnant during this time. Researchers have followed these women and their babies in order to understand the consequences of adverse nutrition in the mothers on the babies' own long-term health.

The results showed that the children in the Dutch Hunger Winter Cohort study grew up to have higher levels of cholesterol in their blood, a risk factor of heart disease and stroke, and had a higher incidence of obesity, diabetes and mental health problems, such as schizophrenia. Followed so far into their 60s, it seems this cohort of individuals has a higher rate of mortality compared with others their age whose mothers were not victims of the famine.

Researchers studied their DNA methylation profiles and found a different pattern in them, too, raising the possibility that epigenetic memory somehow mediated the link between poor nutrition in the womb and later adverse effects upon health. Just how this happens remains a mystery, but one line of enquiry is into the notion that DNA methylation at certain regions of the genome might bring a kind of positive natural selection to the embryo in the face of low nutrition, but it might have an adverse effect upon health later in life.

▲ Baroness Ella van Heemstra (left) is greeted by her daughter, stage and screen star Audrey Hepburn. Ella van Heemstra was Dutch, and the family was caught up in the 1944–45 famine. Audrey had lifelong negative medical repercussions, suffering from anaemia, respiratory illnesses and oedema as an adult.

Results from further studies have added to the growing body of evidence that what we chose to do to our own bodies might be indelibly inked on our epigenetic record and transmitted to some extent to our offspring. Some studies have looked at the possibility that fathers can transmit to their offspring traits that are the result of adverse experiences. For example, some research suggests that children of fathers who have suffered trauma are more likely than others to have non-insulin dependent diabetes and mental health issues. One mechanism cited to explain this is transmission of changes to the RNA, the molecules that translate the genetic code from DNA.

It is still extremely early days in this area of science, though, and it is still not certain how and indeed even if transmission through the epigenetic pathway actually occurs – results of recent studies are conflicting. It remains to be seen whether we can get to the bottom of whether what we do to ourselves today will remain etched upon the epigenetic memory in future generations.

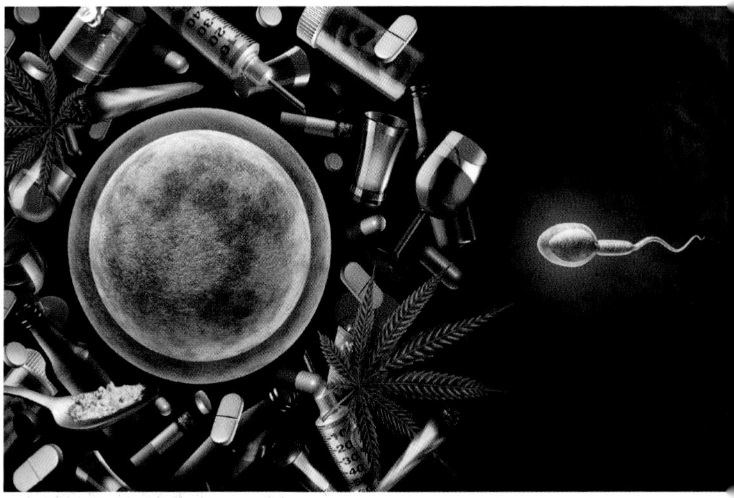

▲ Could the ill effects of our bad habits be transmitted to the next generation through epigenetic changes to our DNA?

The microbiome

Colonising every square millimetre of our bodies – on our skin, in our mouths, in our guts – pretty much everywhere, are billions and billions of microbes. These bacteria, viruses, fungi and archaea (collectively known as microbiota) contain between them somewhere between 2 and 20 million genes, a thousand times more than the 20,000 that constitute our own genetic code. So 99% of the genes found in the body are actually contained within this 'second genome', known as the microbiome. And it appears that the microbiome plays a part in keeping us healthy, but that it is also implicated in causing chronic medical conditions.

How can the microbiome affect health

Ordinarily, we associate microbes with having harmful effects and we try to eradicate them from our mouths and our skin using antibacterial washes, and by blitzing them using antimicrobial medications, such as antibiotics. But it seems that most of the microbes that colonise our bodies, inside and out, are not only 'friendly', but even essential to good general health. For example, they help digest food, bolster the immune system, produce vitamins and fight off disease-causing microbes.

Similarly, throwing the right mixture of these microbes off balance somehow can have adverse effects on our well-being. Evidence links changes to the microbiome with conditions related to obesity, inflammatory diseases of the bowel, allergies and even Parkinson's disease. So, does the microbiome hold the secret to why there may be a higher incidence of some of these medical conditions over recent years? This is a matter for the Human Microbiome Project to unravel further in the future, as findings from this large-scale research study come to light.

Parkinson's disease and the microbiome

Parkinson's disease is a progressive, neurological condition that causes tremors and difficulty in walking. Affected people will ultimately need help to do most day-to-day things, and will eventually lose their ability to walk. Parkinson's disease results from the death of nerve cells in part of the midbrain called the *substantia nigra*. This is where we produce dopamine, a hormone that has a role in facilitating

▼ The nerve cells in Parkinson's disease contain abnormal clumps of protein called Lewy bodies.

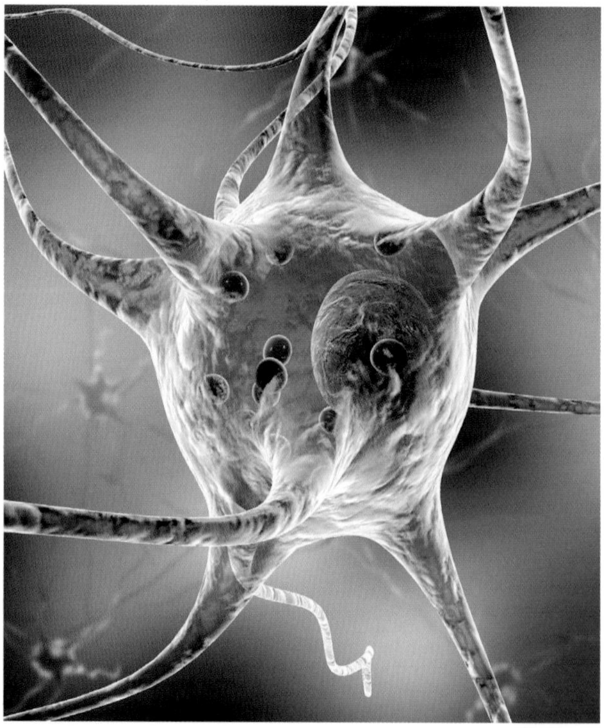

Phylogenetics

Laboratory tests used to work out the composition of our microbiota and how it differs from one person to the next relies upon the detection of ribosomal RNA (rRNA) – the type that bacteria, viruses, fungi and archaea produce. One particular type of rRNA found in microbes, called 16S, contains regions that both are specific to a particular kind of microbe and have not changed in millions of years, very useful in analysing the multitude of bugs that make up each individual's microbiome. This branch of evolutionary microbial science is called phylogenetics, a field that was first conceived in the 1970s, when no one thought of its future potential as a test of our microbiome array. There may well be other niche areas of genetic science gathering dust that we could rediscover and use to expand our knowledge of the role genetics plays in health and disease, and to diagnose and treat medical conditions.

▶ **Microbes colonise our bodies pretty much everywhere, adding a 'second genome' into the mix.**

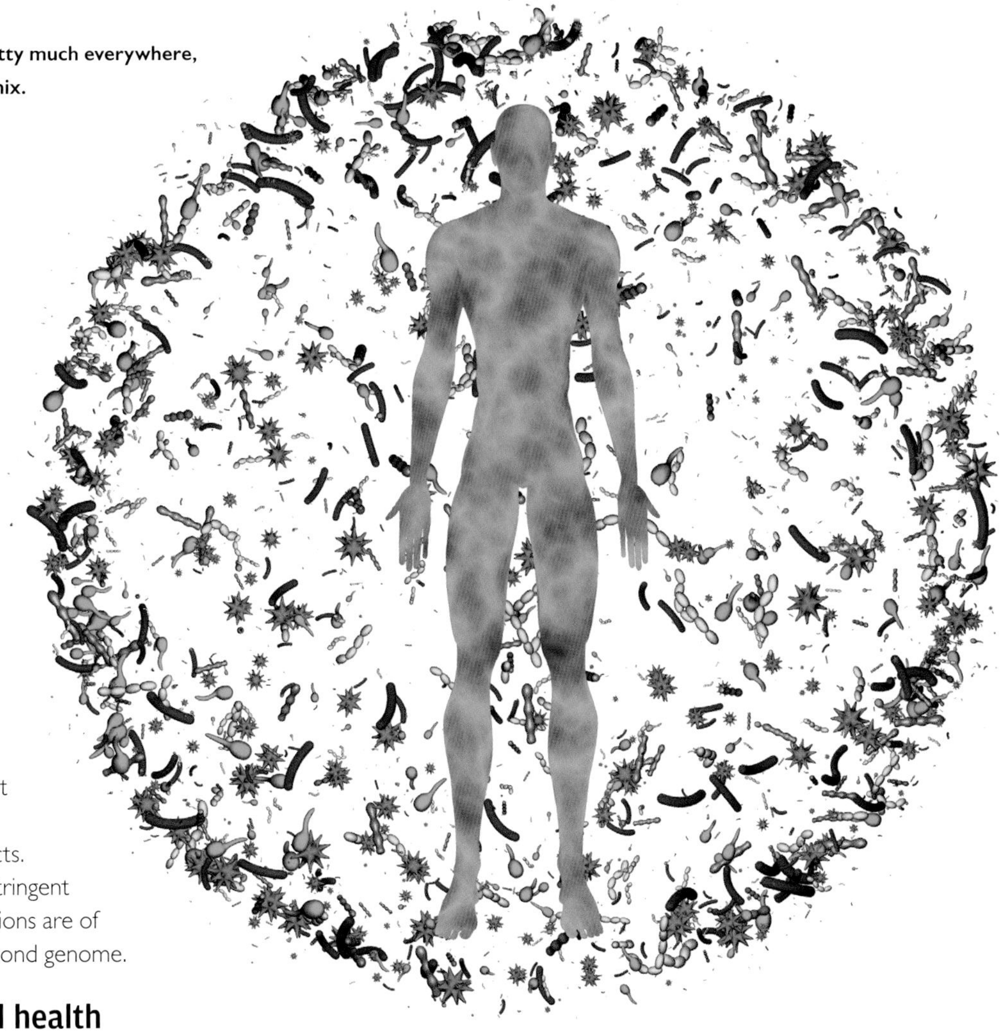

movement, among various other things, such as experiencing emotions and reward sensations, cognition and learning.

The cause of Parkinson's disease is not known, but the answer may lie in the microbiome: the molecule, alpha-synuclein, which forms clumps in nerve cells that are toxic to the brain, is also found in the gut of people affected by Parkinson's disease.

Research using mice has suggested that Parkinson's disease may actually start in the gut, mediated by some constituent of the microbiome producing the molecule that has toxic brain effects. While this theory needs further stringent scientific exploration, early indications are of a possible association with the second genome.

Microbiome and mental health

Through a potential two-way brain–gut axis, researchers have suggested that metabolites produced by gut organisms might reach the brain and affect mental health, and that the brain might then send back messages to the gut to alter the composition of microbes in response. It is definitely early days in this area of microbiome research, but future prospects look interesting.

One recent study demonstrated a link between mental health and the gut microbiome, where indicators of a higher quality of life were positively associated with the presence of two particular regulars in the gut microbiome, *Faecalibacterium* and *Coprococcus*. On the other hand, people with depression have demonstrated significantly less abundant *Coprococcus*, as well as less of another gut bug, *Dialister*. Some gut microbes produce a dopamine metabolite that has been associated with higher quality of life through stimulating the so-called reward system in the brain. So early days it may be, but the future looks promising for improved understanding of a potential role of the microbiome in mental health. More will surely follow.

Obesity and the microbiome

A role for the microbiome in association with obesity is also just starting to emerge. Results of studies in mice and in humans show that the composition of the gastrointestinal microbiome could in some way be related to the reason for an increase in the prevalence of obesity, too.

Possible links between the composition of gut microbiome and Type-2 diabetes, a serious health condition that is more common in people who are obese, are starting to emerge. Scientists are generating profiles of gut microbiota in obese versus lean people to help understand how our gut microbiome might respond to nutritional factors that drive over-eating and therefore obesity. In the future, researchers will look at a possible contribution of gut microbiota in obesity, considering areas such as whether these gut colonisers are releasing any products of microbe metabolism into the blood circulation, driving obesity. Perhaps one day, ways of modulating the gut microbiome might help tackle this and many other modern-day challenges to health and well-being.

Genetic engineering

Genetic engineering is the process of manipulating the genetic code of an organism by altering its basic DNA structure. It involves either inserting a new genetic sequence into the host or removing an existing one; or enhancing or disabling the function of a gene through altering the genetic code. As a result, we can use genetic engineering in as varied ways as medical research, developing new medications (such as vaccines, and enzyme and hormone replacement therapies) and gene therapy. The technologies require stringent regulation on a global scale to ensure we put any advances to good use – to help us, rather than ruin us.

Gene editing

A specific scientific method of genetic engineering, gene editing cuts DNA at a specific, targeted position in the genetic code in order to remove a particular sequence (for example, a stretch containing a disease-causing variant), or to insert a new block, perhaps a section of genetic sequence to replace a faulty gene. The cell's naturally occurring DNA-repair machinery joins up the loose ends, sealing the new sequence into place, fully incorporated into the genome.

We can use gene editing both in the laboratory setting (*in vitro*), as a tool to help conduct medical research and to develop new treatments for genetic disease, and in patients (*in vivo*), where directly administering a gene-editing system might one day be possible to hunt down and treat diseases mediated through a genetic mechanism, such as cancer.

CRISPR-Cas9

CRISPR-Cas9 is a method of editing the genome in a controlled and precise way. It is a find-and-replace gene-editing technique programmed to identify a precise region of the genome, cut it out and insert new, desired DNA sequence. It is composed of two elements. The CRISPR (clustered regularly interspaced short palindromic repeats) component contains a short sequence of RNA that is

Gene-editing embryos

There are very strict guidelines for anyone intending to undertake gene editing on embryos and licenses are very hard to obtain. When an unregulated scientist in China announced in December 2018 that he had edited the genomes of twins Lulu and Nana as day-old embryos, the world was stunned. He claimed to have used CRISPR gene editing in the laboratory to alter the function of a gene called *CCR5* (chemokine, CC motif, receptor 5) with the intention of making the sisters immune to infection with HIV, as their father was HIV-positive. Authorities emphatically voiced concerns that he had opened a Pandora's Box for the genomic era.

He had broken several regulations regarding the use of gene-editing technology in humans. In the UK, the law states that we cannot implant embryos with artificially altered DNA, and (in fact) that scientists must destroy them within 14 days. As a rule, the rest of the world agrees. By acting alone, the Chinese scientist had lost an opportunity for sharing knowledge and undertaken a genetic Russian roulette. After all, the consequences of gene editing are largely unknown and entirely out of our control – who knows what side-effects the changes

might create? For example, although twins are now apparently protected against HIV, the new DNA strand may disrupt the function of another gene to cause serious illness of another kind. If we use genetic editing to remove a faulty gene with an undesirable effect, who's to say that we might inadvertently also disrupt the function of another gene with beneficial effect? Perhaps even one we don't yet know about. Regulation and open debate are essential now and in the future.

The slippery slope?

With rapid advances in genetic technology like gene editing come questions and heated debate over its uses and whether or not they are legitimate and appropriate. Fundamentally, altering the genetic code rightly raises concerns that we are on a steady decline to a race of super-humans, and questions are asked about whether the methodology might be unsafely employed to create so-called designer babies, genetically modified to have attributes in their appearance, personalities, talents and intelligence that some might see as desirable. In reality, though, the specific genetic recipe that confers characteristics like these is not known. As we have already seen, these traits are more likely the result of a whole host of many subtle variants within the genetic code, each with a tiny effect, that when combined and interacting with the environment make us the individuals we are.

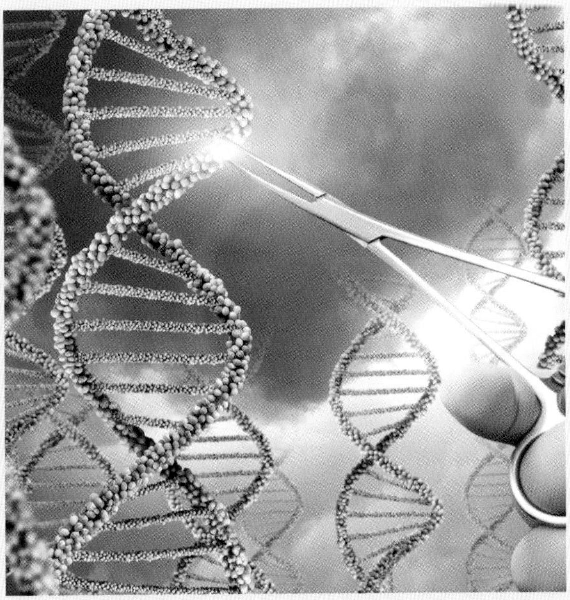

complementary to and a precise match for the target DNA. This enables the component to find the DNA it needs to replace, rather like entering an address into a satellite navigation system. Once accurately located, the nuclease enzyme Cas9 (CRISPR-associated protein 9) acts as a pair of chemical scissors that cut the DNA at the exact location intended. The replacement DNA is then secured in place by the cell's DNA-repair system.

We can use CRISPR-Cas9 technology in medical research to study gene function in disease. This is because it disrupts the normal function of the gene, enabling us to study the consequence. Researchers have also used it to study the early development of the human embryo, which we now know (thanks to this research) is orchestrated by a small number of genes.

Avoiding the dark side

Some people have concerns about a potential dark side to advances in genetic technology and their potential for catastrophic misuse. Established in 1991, the Nuffield Council on Bioethics mediates this kind of open dialogue among scientists, doctors, ethicists, the clergy and members of the general public to advise policy makers and make recommendations to government.

The Council of Europe's 1997 Convention on Human Rights and Biomedicine anticipated early on the potentially far-reaching consequences of DNA-based technologies and states in article 13 of its treaty that 'an intervention seeking to modify the human genome may only be undertaken for preventive, diagnostic or therapeutic purposes and only if its aim is not to introduce any modification in the genome of any descendants'.

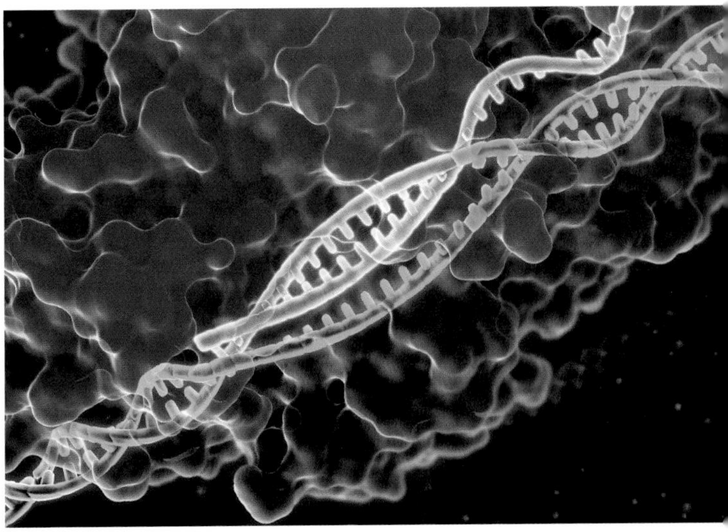

▲ Gene editing using CRISPR-Cas9. Cas9 (purple) attaches to the target DNA using an RNA strand as a guide (orange). Cas9 'locks' onto the DNA and cuts at precisely the right place (green).

Open and transparent discussion in combination with agreed regulation enables the adoption of welcome medical advances that would be beneficial to society, such as IVF (*in-vitro* fertilisation) technology, but also mitigates against implementing possibly damaging developments, at least, in some cases such as gene editing, until we better understand their benefits.

Cloning

Cloning produces organisms that are genetically identical. While we teeter on the brink of understanding the potential for cloning in humans, its use is fraught with difficulty and debate.

Dolly the sheep

Researchers at the Roslin Institute in Scotland, in their bid to produce genetically modified livestock, created Dolly. They took the nucleus of a cell from the mammary gland of a six-year-old Finn Dorset sheep and introduced it into the egg of a Scottish Blackface sheep, that had had its own nucleus removed. They then found another Blackface to act as a surrogate mother.

Dolly was successfully delivered in 1996, live and well.

▲ Dolly the sheep, now stuffed and on display in the National Museum of Scotland.

▼ Technological advances are helping us to unravel the secret of how human life begins.

The biological age of her cells, in spite of her chronological age, seemed older, more in keeping with the age of the sheep from which the cloned mammary gland cell had originated. She had a number of lambs herself and lived to be six years old, dying as a result of a virus that leads to lung cancer in sheep.

Technically it is much harder to conduct cloning in humans – and no one yet has the technology to do it. Unlike with other mammals, in humans spindle proteins, which an embryo needs for cell division after the sperm has fertilised the egg, are found very close to the chromosomes. This means that when scientists try to remove the nucleus from a human, they also remove these proteins, preventing cell division.

Of course, technical difficulties aren't the only thing: there are major ethical concerns surrounding human reproductive cloning, which is, as a result, illegal under international law.

Therapeutic cloning

Therapeutic cloning is different in concept to human reproductive cloning because its aim is to create a treatment for a medical condition using stem cells containing the patient's own genetic code. Stem cells are a particular kind of cell that have the potential to grow into one of many different cell types.

At its early stages in development, an artificially created embryo is a rich source of stem cells. In therapeutic cloning,

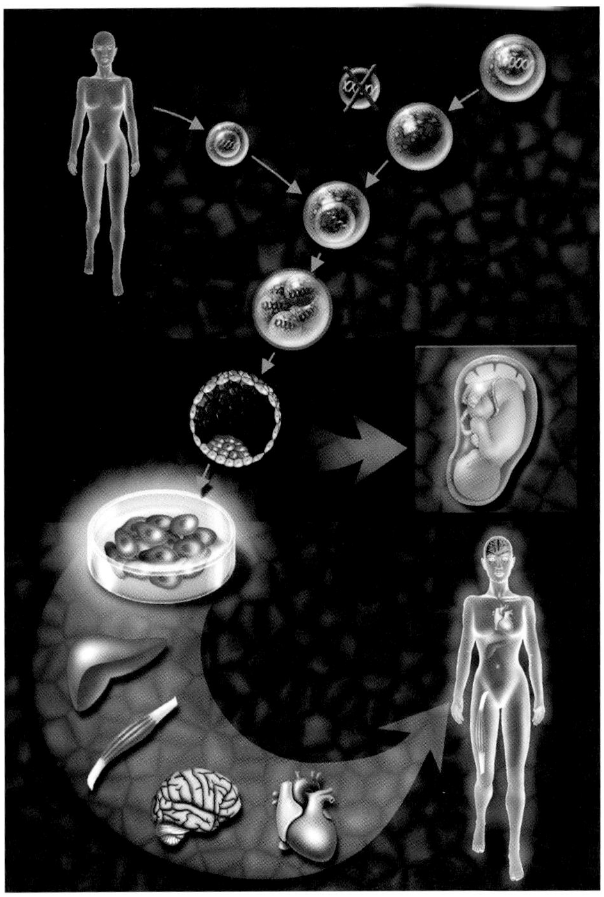

▲ **Therapeutic cloning could one day help us grow spare organs that are genetically compatible for transplantation.**

Whole genome sequencing

Never before have we had such unprecedented access to our genomes. Major technological advances now make it possible for us to sequence the entire genetic code from start to finish in one go – that's 3 million base pairs of DNA. And the time it takes to do this is becoming shorter, too, with the advent of next-generation sequencers that are fast and have a much higher throughput. The price is becoming more affordable as well, in part because so many genomes are being analysed, providing economies of scale. In the near future it might take only 24 hours to have your entire DNA library read at a price you can afford. It might also be the case that newborn babies have the heel prick test to look right across the genome for genetic illnesses that might be treatable, not just a handful of conditions.

However, the rolling out of whole genome sequencing (WGS) on a population level brings significant challenges – both logistical and ethical, which we must address before we can proceed on such a large scale.

Bioinformatics needs to rise to the occasion, so we can make sense of the vast amount of data WGS will generate. We'll need innovative IT solutions to store and handle that data. Linking genomic data to electronic medical records will also be a challenge, but will be necessary if we are to harness the genome in such a way that brings the greatest benefits to our health. It will be of paramount importance to keep people's genomes in a confidential and secure way.

We must take care not to cause any harm through carrying out WGS across the general population. And as a society, we must ensure that we do not create stigmatism or a genetic hierarchy based on test results – the genome has the power to transform medicine and public health in the future, but it must be handled with care.

scientists harvest stem cells from an embryo, remove the nuclei from them and replace them with the nuclei from the patient's cells. The result is a batch of cells with a genetic code identical to the patient's. We can then programme those stem cells to develop into the cells required to treat the patient's condition. Because the cells are identical to the patient's, the immune system does not see them as foreign, so accepts them.

However, this is not without its own technical and ethical issues. Embryonic cells with their potential for growth in some ways have similarities with cancer cells and their potential for growth, raising the concern that therapeutically cloned cells might actually grow out of control and cause cancer. The ethical debate relates to creating embryos with the sole purpose of using them as a source of cells. Once they have served their purpose, the embryos are destroyed. Is this a flagrant disregard for life?

In the future, using immortalised pluripotent stem cells (iPSC) will likely overcome such moral dilemmas. iPSC are derived from adult cells, including a type of blood cell and liver and skin cells obtained from a biopsy sample and reprogrammed to become a stem cell, which scientists can then instruct to differentiate into the desired cell type. Furthermore, umbilical cord blood withdrawn when a baby is delivered contains stems cells that we can use for therapeutic cloning. In fact, there have been early successes using this method to treat blood cancers, such as leukaemia and lymphoma, paving the way for future hopes in this area.

Glossary

Adipose tissue – Loose connective tissue that stores energy in the form of fat, as well as insulating and cushioning the body.

Allele – A variant form of a gene that produces one particular trait.

Amino acid – Short for 'alpha-amino carboxylic acid', an amino acid is an organic molecule made up of carbon, hydrogen, oxygen and nitrogen and a variable R group or side chain that is unique to every amino acid.

Autosomal recessive disorder – A disorder caused by an error in DNA code that occurs in a recessive gene and is passed on by both parents.

Chromosomal aneuploidy – Having a different number of chromosomes from the usual 46.

Chromosomes – Structures found within the cell nucleus that comprise strands of DNA tightly coiled around special proteins called histones.

Codon – A sequence of three DNA or RNA nucleotides that specifies a particular amino acid or a stop or start signal during protein synthesis.

Collagen – The most abundant protein in the human body, collagen is found in connective tissues, muscles, blood vessels, bones, cartilage, tendons and ligament, among other places.

Cytoplasm – A gel-like filling inside a cell, in which are located the various organelles: nucleus, endoplasmic reticulum, Golgi apparatus and mitochondria.

DNA – Deoxyribonucleic acid is the heritable material found in genes. It is a molecule that consists of two strands of nucleotide bases strung along a sugar-phosphate backbone, which are assembled as a double helix.

Endoplasmic reticulum – Found in the form of flattened sacs within the cytoplasm of eukaryotic cells, this continuous membrane system is effectively where proteins are manufactured and packaged.

Epigenetics – The study of the chemical modifications to the DNA and the effects they have on the genome.

Exons – The parts of the nucleotide sequence that encode proteins. Exons amount to just 2% of the genetic code. Collectively they are known as the exome.

Eukaryotic – Usually large cells that have an organised nucleus or organelles with a nuclear envelope.

Fibrils – Supercoiled and entwined fibres.

Fibroblasts – Large, flat, spindle-shaped cells that are the principal active cells of connective tissue.

Genes – Made up of DNA, genes are specific linear sequences of nucleotides located at specific positions on a chromosome that serve as units of heredity.

Gene expression – A process that determines when, where and how much protein is made.

Gene pool – The genetic catalogue of everyone who makes up a particular population.

Genetics – The study of heredity, traits transmitted through the DNA from one generation to the next.

Genetic code – The set of instructions contained within DNA that all living things need to construct the building blocks of life: proteins.

Genetic variation – Random changes in the genetic code.

Golgi apparatus – An organelle found in eukaryotic cells that, like the endoplasmic reticulum, modifies, sorts and packages proteins.

Haemoglobin – A protein that transports oxygen, as well as other chemicals and compounds, around the body.

Heteroplasmy – The presence of more than one type of mitochondrial DNA within a cell or individual.

Heterozygous – Having two different alleles.

Homozygous – Having two alleles that are the same.

Introns – 'Junk' or non-coding sections of a nucleotide sequence that are spliced out before RNA is translated into a protein.

Macromolecule – A very large molecule containing thousands of atoms or more.

Meiosis – Cell division and multiplication that results in the production of four daughter cells, called gametes, that each contain half the number of chromosomes as the mother cell. During the process, two lots of division occur.

Microbiome – Dubbed the 'second genome', this is the collective name for the billions of microbes that inhabit our bodies.

Microsatellites – Very short DNA segments – two to five bases long – that are repeated in tandem and can be used more precisely than minisatellites for DNA profiling.

Minisatellites – Sections of the genetic code where the DNA sequence is repeated multiple times, up to 50. The number of repetitions is unique to every person, except identical siblings, and can be used as a sort of bar code that can identify an individual. It is commonly used in DNA profiling.

Mitochondrion – An organelle within a cell that provides the energy a cell needs to produce maximum output. Cells contain between 1,000 and 2,000 mitochondria, depending on their energy requirements.

Mitosis – Cell division and multiplication that results in the production of two exact replicas of the original cell, containing the same number of chromosomes.

Natural selection – This is when new physical traits that are more advantageous to survival are favoured above all others. There are three types: stabilising, directional and disruptive.

Nucleotide bases – There are four types of bases in DNA: the purines – adenine (A) and guanine (G); and the pyramidines – cytosine (C) and thymine (T). These are found in base pairs, AT and CG, and slot together in a certain way to make a double helix. In RNA, instead of the nucleotide base thymine (T), RNA substitutes uracil (U). The base pairs for RNA are AU and CG.

Obligate carrier – Someone who passes on an inherited pathogenic variant but doesn't necessarily express the condition themselves.

Oncogenes – A group of genes responsible for controlling normal cell growth and division.

Organelle – A tiny structure that performs a specific function within a cell, such as mitochondria, chloroplasts and nuclei. They are often enclosed by their own membrane.

Pathogenic variants – Genetic variants that cause faulty proteins to be constructed, or even stop them from being synthesised, and cause genetic disease.

Polygenic – Involving multiple genes.

Polymerase – An enzyme used to make strands of DNA and RNA.

Polypeptide chain – The basic structure needed to create a mature protein molecule, comprised of amino acids joined together by interlinking peptide bonds.

Prokaryotic cells – Very small cells in which DNA is not organised into a nucleus, and in which there is usually no organelle.

Radiocarbon dating – A technique that allows us to tell the age of ancient objects that contain organic matter by working out how much carbon-14 has decayed.

Reduced penetrance – When someone has a disease-causing genetic variant, but they do not have the condition.

Reproductive fitness – An individual's success at reproducing.

Ribosomes – Tiny particles containing RNA and associated proteins that link amino acids together in the order prescribed by messenger RNA molecules.

RNA – Ribonucleic acid serves, in simplistic terms, to convert the genetic information of DNA into proteins.

SNP – Single nucleotide polymorphism, or 'snips', refers to different forms of the base at a particular position in the genome. SNPs have the ability to influence the efficiency of protein function.

Spliceosomes – Cellular machines that find the start-and-stop sequences inside RNA.

Transcription factors – Proteins that initiate the conversion of DNA into RNA.

Bibliography

Carey, Nessa (2019). *Hacking the Code of Life: How gene editing will rewrite our futures.* Icon Books Ltd

Carey, Nessa (2015). *Junk DNA: A Journey Through the Dark Matter of the Genome.* Icon Books Ltd

Carey, Nessa (2012). *The Epigenetics Revolution: How Modern Biology is Rewriting Our Understanding of Genetics, Disease and Inheritance.* Columbia University Press

Darwin, Charles (2017). *On the Origin of Species.* Macmillan Collector's Library

Dawkins, Richard (2006). *The Selfish Gene.* OUP Oxford

Dawkins, Richard (1986). *The Blind Watchmaker.* Longman

Doudna, Jennifer and Sternberg, Samuel (2017). *A Crack in Creation: The New Power to Control Evolution.* Bodley Head

Epstein, David (2014). *The Sports Gene: Talent, Practice and the Truth About Success.* Yellow Jersey

Firth, Helen and Hurst, Jane (2017). *Oxford Desk Reference: Clinical Genetics and Genomics.* OUP Oxford

Harari, Yuval Noah (2014). *Sapiens: A Brief History of Hum ankind.* Harvill Secker.

Harper, Peter (2017). *Practical Genetic Counselling.* CRC Press

Jones, Steve (1996). *In the Blood: God, Genes and Destiny.* HarperCollins

Jones, Steve (1993). *The Language of the Genes.* HarperCollins

Metzl, Jamie (2019). *Hacking Darwin.* Sourcebooks

Mukherjee, Siddhartha (2016). *The Gene: An Intimate History.* Bodley Head.

Ramakrishnan, Venki (2018). *Gene Machine: The Race to Decipher the Secrets of the Ribosome.* Oneworld Publications

Reich, David (2018). *Who We Are and How We Got Here: Ancient DNA and the new science of the human past.* OUP Oxford

Ridley, Matt (2003). *Genome: The Autobiography of a Species in 23 Chapters.* Us Imports

Ridley, Matt (2000). *Genome: The Autobiography of a Species in 23 Chapters.* HarperCollins

Roberts, Alice (2018). *Evolution: The Human Story.* DK

Roberts, Alice (2014). *The Incredible Unlikeliness of Being: Evolution and the Making of Us.* Heron Books

Russel Wallace, Alfred and Berry, Andrew (2014). *The Malay Archipelago.* Penguin Classics; UK ed.

Rutherford, Adam (2016). *A Brief History of Everyone Who Ever Lived: The Stories in Our Genes.* W&N

Strachan, Andrew (2015). *New Clinical Genetics.* Scion Publishing Ltd.

Turnpenny, Peter and Ellard, Sian (2017). *Emery's Elements of Medical Genetics.* Elsevier

Watson, James; Berry, Andrew and Davies, Kevin (2017). *DNA: The Story of the Genetic Revolution.* Knopf Publishing Group

Index